SOLID WASTE MANAGEMENT

Policy and Planning for a Sustainable Society

SOLID WASTE MANAGEMENT

Policy and Planning for a Sustainable Society

Edited by

Elena Cristina Rada, PhD

APPLE ACADEMIC PRESS

Apple Academic Press Inc.
3333 Mistwell Crescent
Oakville, ON L6L 0A2
Canada

Apple Academic Press Inc.
9 Spinnaker Way
Waretown, NJ 08758
USA

First issued in paperback 2021

Exclusive worldwide distribution by CRC Press, a member of Taylor & Francis Group

ISBN 13: 978-1-77463-622-0 (pbk)
ISBN 13: 978-1-77188-374-0 (hbk)

Library and Archives Canada Cataloguing in Publication

Solid waste management (Oakville, Ont.)
Solid waste management : policy and planning for a sustainable society / edited by Elena Cristina Rada, PhD.

Includes bibliographical references and index.
Issued in print and electronic formats.
ISBN 978-1-77188-374-0 (hardcover).--ISBN 978-1-77188-375-7 (pdf)
1. Refuse and refuse disposal--Environmental aspects. 2. Refuse and refuse disposal--Planning. 3. Refuse and refuse disposal--Case studies. I. Rada, Elena Cristina author, editor II. Title.

TD791.S648 2016 363.72'8 C2015-906894-0 C2015-906895-9

Library of Congress Cataloging-in-Publication Data

Names: Rada, Elena Cristina.
Title: Solid waste management : policy and planning for a sustainable society / [edited by] Elena Cristina Rada, PhD.
Description: Toronto : Apple Academic Press, [2015] | Includes bibliographical references and index.
Identifiers: LCCN 2015039207 | ISBN 9781771883740 (alk. paper)
Subjects: LCSH: Refuse and refuse disposal. | Refuse and refuse disposal--Biodegradation. | Waste minimization.
Classification: LCC TD791 .S65285 2015 | DDC 363.72/85561--dc23
LC record available at http://lccn.loc.gov/2015039207

Apple Academic Press also publishes its books in a variety of electronic formats. Some content that appears in print may not be available in electronic format. For information about Apple Academic Press products, visit our website at **www.appleacademicpress.com** and the CRC Press website at **www.crcpress.com**

ABOUT THE EDITOR

ELENA CRISTINA RADA, PhD

Elena Cristina Rada, PhD, earned her master's degree in Environmental Engineering from the Politehnica University of Bucharest, Romania; she received a PhD in Environmental Engineering and a second PhD in Power Engineering from the University of Trento, Italy, and the Politehnica University of Bucharest. Her post-doc work was in Sanitary Engineering from the University of Trento, Italy. She has been a professor in the Municipal Solid Waste master's program at Politehnica University of Bucharest, and has served on the organizing committees of "Energy Valorization of Sewage Sludge," an international conference held in Rovereto, Italy, and Venice 2010, an International Waste Working Group international conference. She also teaches seminars in the bachelor, master, and doctorate modules in the University of Trento and Padua and Politehnica University of Bucharest (master, doctorate, bachelor modules) and has managed university funds at national and international level. Dr. Rada is a reviewer of international journals, a speaker at many international conferences, and the author or co-author of about a hundred research papers. Her research interests are bio-mechanical municipal solid waste treatments, biological techniques for biomass characterization, environmental and energy balances regarding municipal solid waste, indoor and outdoor pollution (prevention and remediation) and health, and innovative remediation techniques for contaminated sites and streams.

CONTENTS

ACKNOWLEDGMENT AND HOW TO CITE

The editor and publisher thank each of the authors who contributed to this book. The chapters in this book were previously published elsewhere. To cite the work contained in this book and to view the individual permissions, please refer to the citation at the beginning of each chapter. Each chapter was read individually and carefully selected by the editor; the result is a book that provides a multiperspective look at the ways in which international policy and planning intersect with waste management. The chapters included examine the following topics:

Part I: Overview and Analysis

- Chapter 1 indicates that the role of biological treatments should be limited to the processing of organic fraction selectively collected.
- Chapter 2 highlights the relative sizes of ecological footprints among materials and indicates that cities monitoring and planning for reductions in material consumption should be aware of both the quantity and ecological impact per unit of individual materials.
- Chapter 3 points out that waste as a problem—as something not valuable—needs to give way to a general societal awareness that waste is a valuable resource.
-

Part II: Case Studies

- Chapter 4 presents the key elements for the best performance and profitability of municipal solid-waste management in a low-income city.
- Chapter 5 analyzes some of the strengths and deficiencies in the current multiple solid waste management system in Minna, a fast-growing city in north central Nigeria, and proposes feasible solutions.
- Chapter 6 contains a study conducted on the solid-waste management system on the island of Santa Cruz in the Galápagos Archipelago, Ecuador.

Part III: Strategies

- Chapter 7 reviews the salient features of methods of bioremediation, its limitations, and recent developments in solid-waste management through bioremediation.

- Chapter 8 calculates the environmental impact of the six identified biosolids reuse options using a life-cycle assessment approach and a community dialogue mechanism.

Part IV: Policy Planning for the Future

- Chapter 9 suggests and discusses policy instruments that could lead toward more sustainable waste management, along with proposals for further development.
- Chapter 10 proposes a research framework of zero-waste management and strategies for low-carbon residential precincts.

LIST OF CONTRIBUTORS

Segun Emmanuel Adebayo
Department of Agricultural and Bioresources Engineering, Federal University of Technology, P.M.B. 65, Minna, Nigeria

Peter Aderemi Adeoye
Department of Biological and Agricultural Engineering, Universiti Putra, 43400 UPM Serdang, Malaysia

Rick Arneil D. Arancon
Department of Chemistry, School of Science and Engineering, Ateneo de Manila University, Quezon City, Philippines; School of Energy and Environment, City University of Hong Kong, Hong Kong

Yevgeniya Arushanyan
KTH Royal Institute of Technology, School of Architecture and Built Environment, Department of Urban Planning and Environment, Division of Environmental Strategies Research, SE-100 44 Stockholm, Sweden

Mattias Bisaillon
Profu AB, Årstaängsvägen 1A, SE-117 43 Stockholm, Sweden

Anna Björklund
KTH Royal Institute of Technology, School of Architecture and Built Environment, Department of Urban Planning and Environment, Division of Environmental Strategies Research, SE-100 44 Stockholm, Sweden

Riccardo Catellani
Department of Civil, Environmental and Mechanical Engineering, University of Trento, Via Mesiano 77, Trento I-38123, Italy

King Ming Chan
Environmental Science Program, School of Life Sciences, Chinese University of Hong Kong, Hong Kong

Emil Cohen
Independent Expert for Waste to Energy Projects, Moshav Aniam, Israel.

Gregory Yom Din
The Open University of Israel, Raanana, Israel; Tel-Aviv University, Tel-Aviv, Israel

John Devlin
PhD Candidate, sd+b Centre, University of South Australia, Australia

Tomas Ekvall
IVL Swedish Environmental Research Institute, PO Box 530 21, SE-400 14 Stockholm, Sweden

Ola Eriksson
Department of Building, Energy and Environmental Engineering, Faculty of Engineering and Sustainable Development, University of Gävle, SE-800 76, Gävle, Sweden

Göran Finnveden
KTH Royal Institute of Technology, School of Architecture and Built Environment, Department of Urban Planning and Environment, Division of Environmental Strategies Research, SE-100 44 Stockholm, Sweden

Tomas Forsfält
Konjunkturinstitutet, P.O. Box 3116, SE-103 62 Stockholm, Sweden

Tiwari Garima
School of Energy and Environmental studies, Devi Ahilya University Indore, India

Mona Guath
KTH Royal Institute of Technology, School of Architecture and Built Environment, Department of Urban Planning and Environment, Division of Environmental Strategies Research, SE-100 44 Stockholm, Sweden

Greger Henriksson
KTH Royal Institute of Technology, School of Architecture and Built Environment, Department of Urban Planning and Environment, Division of Environmental Strategies Research, SE-100 44 Stockholm, Sweden

Meidad Kissinger
Department of Geography and Environmental Development, Ben-Gurion University of the Negev, Beer-Sheva, 8410501, Israel

Tsz Him Kwan
Environmental Science Program, School of Life Sciences, Chinese University of Hong Kong, Hong Kong

Elisabeth (Lisa) R. Langer
Scion (New Zealand Forestry Research), PO Box 29 237, Christchurch, 8540, New Zealand

Alan C. Leckie
Scion (New Zealand Forestry Research), PO Box 29 237, Christchurch, 8540, New Zealand

Steffen Lehmann
Chair Professor of Sustainable Design; Director, sd+b Centre and CAC-SUD Centre, University of South Australia, Australia

Carol Sze Ki Lin
School of Energy and Environment, City University of Hong Kong, Hong Kong

Rafael Luque
Departamento de Química Orgánica, Universidad de Córdoba, Campus Universitario de Rabanales, Córdoba, Spain; Department of Chemical and Biomolecular Engineering (CBME), Hong Kong University of Science and Technology, Kowloon, Hong Kong

James E. McDevitt
Scion (New Zealand Forestry Research), PO Box 10 345, The Terrace, Wellington, 6143, New Zealand

Jennie Moore
School of Community and Regional Planning, University of British Columbia, Vancouver, V6T 1Z2, Canada

John Jiya Musa
Department of Agricultural and Bioresources Engineering, Federal University of Technology, P.M.B. 65, Minna, Nigeria

Ulrika Gunnarsson Östling
KTH Royal Institute of Technology, School of Architecture and Built Environment, Department of Urban Planning and Environment, Division of Environmental Strategies Research, SE-100 44 Stockholm, Sweden

Queena K Qian
Endeavour Australian Cheung Kong Post-Doc Fellow, sd+b Centre, University of South Australia, Australia

Elena Cristina Rada
Department of Civil, Environmental and Mechanical Engineering, University of Trento, Via Mesiano 77, Trento I-38123; Department of Biotechnologies and Life Sciences, University of Insubria, Via G.B. Vico 46, Varese I-21100, Italy

Marco Ragazzi
Department of Civil, Environmental and Mechanical Engineering, University of Trento, Via Mesiano 77, Trento I-38123

William E. Rees
School of Community and Regional Planning, University of British Columbia, Vancouver, V6T 1Z2, Canada

Mohammed Abubakar Sadeeq
Department of Agricultural and Bioresources Engineering, Federal University of Technology, P.M.B. 65, Minna, Nigeria

Jenny Sahlin
Profu AB, Götaforsliden 13, SE-43134 Mölndal, Sweden

Xavier Salazar-Valenzuela
Office of Environmental Management, Municipality of Santa Cruz, Av. Charles Darwin y 12 de Febrero, Puerto Ayora 200350, Santa Cruz Island, Galapagos, Ecuador; Central University of Ecuador, Galapagos Head, Barrio Miraflores entre Petrel y San Cristóbal, Puerto Ayora 200350, Santa Cruz Island, Galapagos, Ecuador

S. P. Singh
School of Energy and Environmental studies, Devi Ahilya University Indore, India

Patrik Söderholm
Luleå University of Technology, Economics Unit, SE-971 87 Luleå, Sweden

Maria Ljunggren Söderman
IVL Swedish Environmental Research Institute, PO Box 530 21, SE-400 14 Stockholm, Sweden; Chalmers University of Technology, Environmental Systems Analysis, Energy and Environment, SE-412 96 Göteborg, Sweden

Åsa Stenmarck
IVL Swedish Environmental Research Institute, P.O. Box 210 60, SE-100 31 Stockholm, Sweden

Johan Sundberg
Profu AB, Götaforsliden 13, SE-43134 Mölndal, Sweden

Jan-Olov Sundqvist
IVL Swedish Environmental Research Institute, P.O. Box 210 60, SE-100 31 Stockholm, Sweden

Cornelia Sussman
School of Community and Regional Planning, University of British Columbia, Vancouver, V6T 1Z2, Canada

Åsa Svenfelt
KTH Royal Institute of Technology, School of Architecture and Built Environment, Department of Urban Planning and Environment, Division of Environmental Strategies Research, SE-100 44 Stockholm, Sweden

Vincenzo Torretta
Department of Biotechnologies and Life Sciences, University of Insubria, Via G.B. Vico 46, Varese I-21100, Italy

Atiq Uz Zaman
PhD Candidate, sd+b Centre, University of South Australia, Australia

INTRODUCTION

Because solid waste is generated everywhere, addressing the environmentally safe management of solid waste is an international challenge. Management strategies vary by country and region, although most programs address waste issues with models consisting of some combination of source reduction, combustion, recycling, and landfills. Per capita waste generation rates also vary significantly by country; factors contributing to such discrepancies include lifestyles and economic structures.

The environmentally safe management of municipal solid waste may always be an issue, simply because societies will continue to generate trash due to increasing populations and the growing demands of modern society. Working together, governments, industry, and citizens have made substantial progress in effectively responding to solid waste issues through source reduction, recycling, combustion, and landfill programs. Such community-tailored programs provide possible long-term solutions to solid waste management.

In chapter 1, the authors discuss a modern integrated management of municipal solid waste that must be developed balancing the relationship between selective collection of recyclable fractions and the residual waste. A perfect system should give streams of selective collection with no impurities and a stream of residual municipal solid waste with only non-recyclable materials. In the real world, the dynamics of the efficiency of selective collection has drastically changed the characteristics of the residual waste in a few areas. This change has consequences from the planning, design, environmental, energy, and economic points of view. The environmental consequences refer both to local impacts and to global ones (greenhouse gases). Chapter 1 deals with these aspects, also referring to the results of two case studies that help to illustrate the advantages and disadvantages of this situation. The first case study refers to an area where the choice of the treatment option for residual municipal solid waste is made after reaching high efficiency of selective collection. The second case study involves a comparison between waste-to-energy plant and a modern landfill, taking into account the role of carbon sink, where the residual waste characteristics are strongly affected by the high efficiency of selective collection.

Ecological footprint analysis (EFA) can be used by cities to account for their on-going demands on global renewable resources. To date, EFA has not been fully implemented as an urban policy and planning tool in part due to limitations of local data availability. In chapter 2, the authors focus on the material consumption component of the urban ecological footprint and identify the solid-waste life-cycle assessment (LCA) approach as one that overcomes data limitations by using data many cities regularly collect: municipal, solid waste composition data, which serves as a proxy for material consumption. The approach requires energy use and/or carbon dioxide emissions data from process LCA studies as well as agricultural and forest land data for calculation of a material's ecological footprint conversion value. The authors reviewed the process LCA literature for twelve materials commonly consumed in cities and determined ecological footprint conversion values for each. The authors found a limited number of original LCA studies but were able to generate a range of values for each material. Their set of values highlights the importance for cities to identify both the quantities consumed and per-unit production impacts of a material. Some materials like textiles and aluminum have high ecological footprints but make up relatively smaller proportions of urban waste streams than products like paper and diapers. Local government use of the solid-waste LCA approach helps to clearly identify the ecological loads associated with the waste they manage on behalf of their residents. This direct connection can be used to communicate to citizens about stewardship, recycling, and ecologically responsible consumption choices that contribute to urban sustainability.

Increasingly tighter regulations regarding organic waste, and the demand for renewable chemicals and fuels, are pushing the manufacturing industry toward higher sustainability to improve cost effectiveness and meet customers' demand. Food waste valorization is one of the current research areas that have attracted a great deal of attention over the past few years as a potential alternative to the disposal of a wide range of residues in landfill sites. In particular, the development of environmentally sound and innovative strategies to process such waste is an area of increasing importance in our current society. Landfill, incineration, and composting are common, mature technologies for waste disposal. However, they are not satisfactory for treating organic waste due to the generation of toxic methane gas and bad odor, high-energy consumption, and slow reaction kinetics. In fact, research efforts have also been oriented on novel technologies to decompose organic waste. However, no valuable product is generated from the decomposition process. Instead of disposing and decomposing food waste, recent research has focused on its utilization as energy source (e.g., for bioethanol and biodiesel production). Organic waste is also useful to generate useful organic chemicals via biorefinery or white biotechnology (e.g., succinic acid and/or bio-plastics). Chapter 3 is aimed to summarize recent development of waste valorization strategies for the sustainable production of chemicals, materials, and fuels through the development of green production strategies. Chapter 3 also provides key insights into recent legislation on management of waste worldwide as well as two relevant case studies (the transformation of corncob residues into functionalized biomass-derived carbonaceous solid acids and their utilization in the production of biodiesel-like biofuels from waste oils in Philippines, as well as the development of a bakery waste based biorefinery for succinic acid and bioplastic production in Hong Kong) to illustrate the enormous potential of biowaste valorization for a more sustainable society. Future research directions and possible sustainable approaches are also discussed with their respective proofs of concept.

The purpose of chapter 4 is to present the key elements for best performance and profitability of municipal solid waste (MSW) management in a low-income city. The research provides an overview of methods and models for integrated planning of a two-phase program: MSW collection and transportation, and MSW treatment. The authors present the case study of Matadi (the Democratic Republic of Congo) that has a low level of the MSW management compared to other African cities. They develop a spreadsheet model for collection and transportation of MSW that is relevant for low-income cities and enables determining the waste collection fee. A CDM decay model is used to predict the GHG emissions in disposal site. The MSW treatment plant in case of Matadi is evaluated; the anaerobic digestion technology selected as appropriate for this plant, the key factors that ensure profitability of the plant are as follows: tipping fee from the municipality (19% of total revenue), amount of carbon credits that can sum up to 16% of the total revenue, expansion of waste collection range from 25 to 50 km. The methods of chapter 4's study can be used for solving waste problem in other low-income cities where the budget for municipal services is scanty, particularly when starting from a very low level of MSW management.

Chapter 5 presents a brief trend in Minna MSW management. Municipal solid waste is a major environmental problem in Minna as in many developing towns. Although strict regulations on the management of solid waste are in place, primitive disposal methods such as open dumping and discharge into surface water are still being used in various parts of the town. Chapter 5 also presents the MSW management structure together with the present situation of generation composition, collection, disposal, and treatment. The chapter also provides a brief discussion on the future challenges. Waste generation increases by more than 200% within two decades and increase in budgetary allocation is not proportional to the waste generation increase. The waste management agency did not have enough resources to tackle the ever-increasing municipal solid waste. The collection process is deficient in terms of manpower and vehicle availability. Bin capacity provided is inadequate, and their locations were found to be inappropriate, thus contributing to the inefficiency of the system. At this time, no treatment is provided to the waste after collection. Underestimation of waste generation rates and inadequate management and technical skills were also responsible for inappropriate waste management system in the town

Chapter 6 analyzes some aspects of the management of municipal solid waste in one of the islands of the Galapagos archipelago. The chapter's aim is to point out a few aspects of an interesting experience that could help decision managers faced with the organization of the waste sector in similar realities. The relevance of this case study consists in the presence of a very famous national park surrounding the inhabited area. The role of tourism in the generation of waste is analyzed too.

Bioremediation is an alternative way to manage or to degrade the waste. It is eco-friendly and much cost effective as compared to other traditional technique such as incineration. The main purpose of chapter 7 is to pay more attention towards bioremediation. The chapter outlines the different processes of bioremediation, their limitation, and the process to remove different heavy metals, and other waste that is harmful to human beings. When metals are treated with microbes, they get accumulated or attached on microbial membrane. And after that, they can be extract from microbes through cell fragmentation.

Chapter 8 reports a life-cycle assessment undertaken to assess the environmental impact of a range of biosolid reuse options selected by the Kaikōura community. The reuse options were identified as: vermiculture and open-air composting; mixture with biochar; direct land application to disturbed sites for forestry using native tree species; and application to exotic forestry plantations or pastoral farmland. The aim of the study was to calculate the possible environmental impacts of the reuse options so the information can be used in a community dialogue process where the fate of the biosolids is decided. All reuse options showed improved environmental performance relative to landfilling. The direct application to land options showed the least environmental impact, and the composting options had the most environmental impact. The authors indicate that this is the first time this approach has been applied to biosolids management in New Zealand, and while there are limitations, the approach should be encouraged in other communities because it increases the engagement of the community with waste management decision-making and the environment

The aim of chapter 9 is to suggest and discuss policy instruments that could lead towards a more sustainable waste management. The paper is based on evaluations from a large-scale multidisciplinary

Swedish research program. The evaluations focus on environmental and economic impacts as well as social acceptance. The focus is on the Swedish waste management system but the results should be relevant also for other countries. Through the assessments and lessons learned during the research program, the authors conclude that several policy instruments can be effective and possible to implement. In particular, they put forward the following policy instruments: "information"; "compulsory recycling of recyclable materials"; "weight-based waste fee in combination with information and developed recycling systems"; "mandatory labeling of products containing hazardous chemicals"; "advertisements on request only and other waste minimization measures"; and "differentiated VAT and subsidies for some services." Compulsory recycling of recyclable materials is the policy instrument that has the largest potential for decreasing the environmental impacts with the configurations studied. The effects of the other policy instruments studied may be more limited, and they typically need to be implemented in combination in order to have more significant impacts. Furthermore, policymakers need to take into account market and international aspects when implementing new instruments. In the more long-term perspective, the chapter's set of policy instruments may also need to be complemented with more transformational policy instruments that can significantly decrease the generation of waste.

The consumption-driven society today produces an enormous amount of waste, which puts pressures on land, pollutes the environment and creates economic burden. "Zero-waste" concept, a whole system approach aiming to achieve no waste along the materials flow through society, has become one of the most visionary concepts for tackling growing waste problems. In chapter 10, system dynamics (SD) approach is applied in the proposed framework for designing the waste management in a zero-waste residential precinct. A cost-benefit analysis (CBA) is incorporated to supplement the SD framework to evaluate the total cost and benefit of waste and resources throughout the material flow chain. The authors propose a list of parameters under the categories of process, technology and infrastructure, socioeconomic and institutional, and social-environment, to be tested in future case study of Bowden village in Australia. Their framework provides an inventory of leverage points to help policymakers design waste policies and allocate resources effectively, with minimum environmental impact and optimum social benefits. It also helps planning professionals and business stakeholders better understand the costs and benefits of different scenarios for achieving a zero-waste residential precinct.

PART I

OVERVIEW AND ANALYSIS

CHAPTER 1

How High Efficiency Selective Collection Affects the Management of Residual MSW

M. RAGAZZI AND E. C. RADA

1.1 INTRODUCTION

In agreement with the European Union directives, a modern integrated management of municipal solid waste (MSW) must ensure a balanced co-existence of selective collection and residual MSW. From the theoretical point of view, a perfect system will give streams of selective collection with no impurities and a stream of residual MSW with only recyclable materials. An important topic is the role of the organic fraction: its selective collection can significantly decrease its content in the residual waste with consequences on the MSW management. More in general, the dynamics of the efficiency of selective collection has drastically changed the characteristics of the residual MSW in a few areas. Today, where the management of MSW already comply with the latest European Union targets, the residual MSW is composed mainly of non-recyclable materials.

These situations are limited to small towns in few regions as the efficiency of selective collection is affected by the verticality of the urbanization.

In this paper, consequences of this unsteady scenario on planning, designing and management are discussed. Additionally a case study is presented referring to real data from an area where selective collection is reaching 65% of efficiency (and more). The most important consequence is related to the significant reduction of recyclable materials in the residual MSW. The analysis of data point out the different efficiency of source separation depending on the curbside system applied (mono-familiar houses versus multi-familiar buildings).

The treatments that are considered for the management of residual MSW in areas with high efficiency selective collection are: combustion, gasification, pyrolysis, integrated thermal plants, aerobic mechanical-biological treatments, anaerobic mechanical-biological treatments and other types of treatment. The considerations related to their viability in the studied scenario concern environmental, energy and economical aspects.

Another important topic that is dealt with in this paper regards a comparative analysis of CO_2eq emissions from waste-to-energy plants and from landfilling of the same residual MSW.

1.2 PLANNING EFFECTS OF HIGH EFFICIENCY SELECTIVE COLLECTION

The dynamics of the residual MSW contents must be taken into account from decision makers for a sustainable planning. A question discussed at the moment concerns the role of bio-mechanical treatments (BMT, in particular with a role of pre-treatment) in regions where strategies for strongly increasing the efficiency for selective collection are implemented. In case of high efficiency of organic fraction of MSW selective collection an existing BMT could not have the necessary amount of substances for a good development of the process. Moreover when a Waste to Energy plant already exists in such regions, the lower heating value could change moving out from the guaranteed range of operation. As an example, in Figure 1 two graphs are shown representing the dynamics of the LHV of bio-dried material and of Refuse Derived Fuel (RDF) (Rada et al., 2009). RDF is produced through separation of inert fractions in two contexts. The first RDF concerns the case of a reality where almost no selective collection is implemented and that includes a significant presence of organic matter in residual MSW (as examples, the latest entries in the European Union—Romania and Bulgaria—and the southern Italy nowadays). The second RDF concerns the case of a region were the selective collection is very enhanced, in particular in terms of organic fraction collection (as example, some sub-regions in the North-East of Europe). The decrease of organic matter down to values below 10% of residual MSW, as a result of extreme selective collection of organic fraction, makes bio-drying unviable, as the core of this process is the availability of putrescible matter that acts as an "engine": In this scenario the energy balance is rather modest, so modest that the process is useless. Similar problems could emerge when the two-stream scenario, based on bio-stabilization of an under-sieve, is proposed: the screening of MSW with a low organic content would have a low efficiency.

Coming back to direct combustion of residual MSW, this option has recently shown a new and attractive way, regarding a "zero landfill" strategy: slag can be completely recycled and fly-ash can be vitrified (with additives) and then used in industrial processes. In this frame, an interesting example can be seen in Noceto, Italy (Schione 2008). In this plant, from 30,000 t y^{-1} of slag, 25,000 t of material for production of concrete, 300 t of non-ferrous metals and 1,500 t of ferrous metals are separated. Concerning fly-ash, its composition is suitable for a vetrification with additives. A variation of this strategy can be based on the separation of the fine fraction of slag to be used with a role of additives for fly-ash co-vetrification. In such a scenario a question could be put: why separating at the source some fractions if a post-treatment could handle them? The answer to this question is not the subject of the present paper.

The increase of putrescible streams separated at the source creates an growing necessity of composting plants (also integrated with anaerobic digestion). From the planning point of view this aspect must be

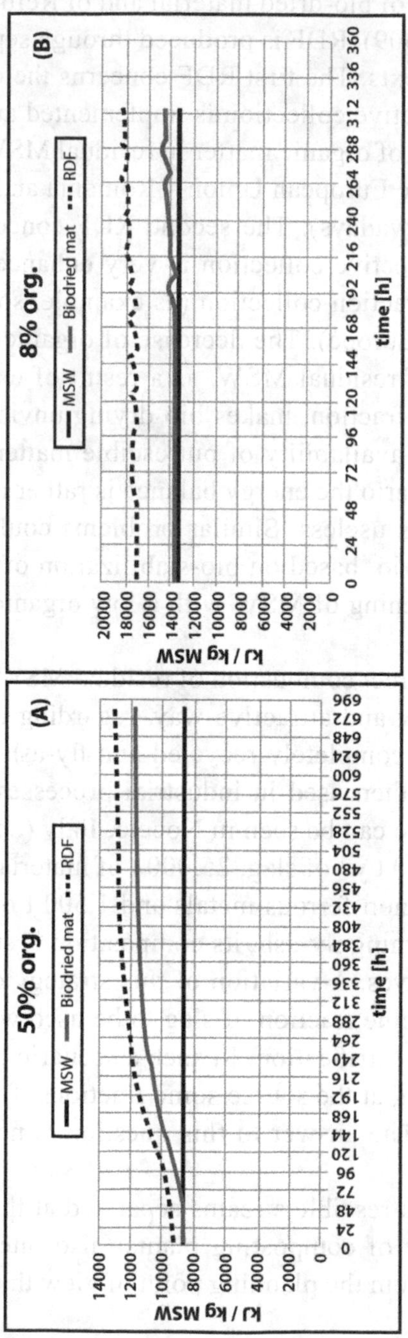

Figure 1. LHV dynamics during the bio-drying process.

managed carefully as an underestimation of the streams to be treated can cause an overloading of the plants with critical operating conditions: the risk of odor events during the operation would be highly probable.

1.3 DESIGNING AND MANAGEMENT

In a scenario that foresees a significant variation of the selective collection efficiencies, the designing of a treatment plant of residual MSW must be developed carefully. For example, the capacity of a thermal treatment plant must be chosen taking into account the energy content of the stream of residual MSW at the opening of the plant and during its operation (15-20 years). If selective collection has not reached a steady state, there is the risk of an over/under dimensioning. The over dimensioning creates diseconomies, instead the under dimensioning can cause emergencies due to the fact that the treatment capacity is not sufficient. A solution to this problem could be the adoption of a modular installations, following the dynamics of the residual MSW stream. Unfortunately an excessive modularity can create some problems on the efficiency of energy recovery and an increase of costs. The designer must find an adequate balancing between all these aspects, but generally a good flexibility is viable only for large plants.

Similar problems, concerning waste quantity and dimensioning, could be found also in the case of a mechanical-biological treatment plant adoption. In this field, anyway, the construction of bio-reactors in modules is a common solution.

Regarding the management aspects of existing plants, it is better to check their flexibility before deciding significant changes in selective collection strategies. For example, if in an area a thermal treatment plant of residual MSW already exists, the sudden activation of organic fraction selective collection may put in difficulty the operator of the plant. Indeed, these plants are designed to operate at a fixed range of LHV. The separation at the source of wet fraction, as the organic one, can raise the LHV of residual MSW up to incompatible values for the plant. Modern plants on the contrary typically can handle most of the situation encountered in practice.

1.4 CASE-STUDY 1—CHOICE OF A NEW TREATMENT PLANT

1.4.1 DEFINITION

The first case-study refers to real data from an Italian area where selective collection is already around 65% of efficiency. The most important consequences are related to the significant reduction of recyclable materials in the residual MSW and to the increase of its LHV that has reached 13 MJ kg^{-1}. The analysis of data pointed out that there is a different efficiency of source separation depending on the curbside system applied: mono-familiar houses guarantee a better selection compared to multi-familiar buildings. That can be explained by the different risk of fine for the users: a common idea is that an incorrect separation cannot be fined when the container refers to a multi-familiar user; in reality the consequence is a collective fine that causes trouble in the multi-familiar community as the responsible is difficult to be found.

In Table 1 some details on the selective collection of this case study are presented. A deeper analysis is useful to understand if the obtained results can be improved: two thirds of the residual waste is composed by recyclable materials. Anyway it must be pointed out that the impurities in the materials separated at the source are not zero. In the case-study a perfect management of the streams of recyclable materials would give a residual MSW stream of 15%. Of course it is not reasonable to plan a management system without flexibility as the citizen is not a "scientist" who separate materials.

The following considerations will take into account a scenario with a selective collection of 65%, a LHV of the residual MSW of 13 MJ kg^{-1} and a percentage of organic fraction in the residual MSW equal to 19% (as resulted from a real scale characterization). The following paragraphs concerns the critical analysis of the viability of some options available for the treatment of the residual MSW.

Table 1. Details of selective collection for the case-study.

Fraction	Curbside collection [kg ab^{-1} y^{-1}]
Light packaging	23
Glass	25
Paper	51
Organic fraction	104
Hand out at the material recycling center	80
Bulky waste	12
Total selective collection	283
Bulky waste non selectively collected	12
Residual MSW	94
Streets management waste	17
Residual MSW recycling center	17
Total residual waste	**140**
Total municipal solid waste	**423**
% selective collection	**67%**

1.4.2 COMBUSTION

The trend of combustors is towards systems able to treat residual MSW with a high lower heating value (LHV). To this concern in recent years the water cooled grate has been developed. Indeed the latest design value of LHV referred to residual MSW collected in areas with high selective collection in Italy is assumed around 13 MJ kg^{-1} (coherent with the above data) with peaks of LHV expected during operation that can reach 20 MJ kg^{-1}. This value can be critical for a conventional grate system (air cooled). The percentage of slag is going down to 20% of the input whilst twenty years ago this value was 30%. In the presented scenario a modern combustor seems to be viable. From the economical point of view it must be put adequate attention to the capacity of the plant referred to the region to be serviced as the conventional combustion could have some unfavorable scale effects when the amount of waste is low: a deep economical analysis can clarify this aspect case by case. Additionally it must be pointed out that a comparison in terms of scale effect should not be made considering the tons to be treated yearly: the comparison must be made in terms of thermal power at the input. Just to clarify this aspect, the thermal power of a plant treating 100,000 t y^{-1} with a LHV as 12 MJ kg^{-1} is the same of a plant treating 150,000 t y^{-1} with a LHV as 8 MJ kg^{-1}.

1.4.3 GASIFICATION, PYROLYSIS AND INTEGRATED SYSTEMS

Gasification and pyrolysis are often proposed as alternative to MSW combustion. A growing interest concern the experience developed in Japan in recent years (Ghezzi, 2008). Anyway it must be pointed out that the combustion systems have been fully commercial for many decades and have accumulated a very wide experience, instead the gasification and pyrolysis are not fully established technologies. The aim of these options is to improve the energy and the environmental balance compared to conventional solutions. The different logics, compared to the European targets, concern the value given to the generated electricity (low in Japan) and to the containment of space (a priority in Japan). The LHV of the residual MSW is viable for these processes. Anyway persists the problem of the heterogeneity of the waste: in some configurations a significant pre-treatment is proposed before entering the pyrolysis/gasification reactor. This pre-treatment can affect the energy balance and the costs of this strategy.

There are different kinds of combined processes, as gasification + combustion or pyrolysis + gasification. In practice gasification coupled with post-combustion is very similar to direct combustion. An advantage could be related to the possibility of a good control of the combustion conditions of a gas and not a solid fuel (MSW).

Finally, the residual MSW of this case-study is viable for alternative thermal processes. From the practical point of view it is important to check the reliability of the plant proposed.

1.4.4 BIOLOGICAL TREATMENT PLANTS

Concerning aerobic mechanical-biological treatments, it can be noticed a comparison between bio-drying (usually single stream) and bio-stabilization (double stream) (Rada et al., 2005a; Rada et al., 2006). The distinction between single or double stream depends on the absence or less of a system of screening upstream. In the two stream system the screening divides the waste in "dry" and "wet" streams. The wet stream is converted into Stabilized Organic Fraction (SOF).

In terms of process, the bio-drying (short process without water addition) aims to evaporate the highest amount of humidity of residual MSW with the lowest consumption of biodegradable volatile solid. For a residual MSW with 19% of organic fraction, during the bio-drying process the volatile solids (VS) consumption could be is 12 gVS kg^{-1}MSW; the weight loss could be about 13% (Rada 2005b). This result can be obtained only with an electrical consumption: this consumption could reach 50 kWh t^{-1}MSW (Rada 2005b). A small energy loss (2%-3%), referred to the initial LHV characterizes this process (Rada 2005b). After a mechanical post-treatment the obtained material (biodried material) can be converted into Refuse Derived Fuel (RDF) and the overall electrical consumption can reach 80 kWh t^{-1}MSW (Rada 2005b). The main result is a concentration of the initial available energy in a lower mass. As shown above, the weight loss for the residual MSW of the case-study is not significant. The only bio-drying could not guarantee reaching the target of 15 MJ kg^{-1}, value that can be considered a threshold for RDF. Indeed assuming 13 MJ kg^{-1} as initial LHV, taking into account the weight loss and the overall energy loss, a value of 14.5 MJ kg^{-1} can be assessed for the bio-dried material. A post-selection of the bio-dried material can extract glass, metals and inert from the residual MSW. In the present case the percentage of these fractions in the residual MSW is about 10%. As a consequence the LHV of the obtainable RDF could be 14.5/0.9 = 16 MJ kg^{-1} of RDF. This result has sense if a RDF based strategy is considerable more viable than direct combustion as it must be pointed out that the LHV of residual MSW is adequately high for a good combustion.

The bio-stabilization (long process with water addition) consumes the highest amount of putrescible volatile solids (those rapidly and medium biodegradable). This consumption depends strongly on the lasting of the process. The bio-stabilization has been always seen as a pre-treatment for landfilling, that becomes mandatory according to the latest regulations. Anyway the new regulations take into account both the LHV and the respirometric index for the pretreated waste (parameter which gives information on the residual biodegradability of the pretreated material). If the bio-stabilization is inserted in a separate stream facility, a parallel production of RDF will exist. It must be taken into account also that a complete bio-stabilization requires energy consumption and processing time longer than bio-drying. The risk of adopting bio-stabilization in a two-stream plant for the case-study is to generate an under-sieve poor in organic matter. Apart from that this strategy has the disadvantage of generating a stream of stabilized organic fraction that in some circumstances can give a problem of sustainability.

Concerning anaerobic mechanical-biological treatment, this solution is less widespread than the aerobic ones in the field of residual MSW treatment. Maybe the reason concerns the difficulty of management caused by the presence of non-biodegradable materials in the residual MSW. The residual MSW of this case study is poor in organic fraction and then its treatment cannot guarantee the extraction of a stream of good quality to be partially converted in biogas.

1.4.4 COMPARISON

Summing up, if a strategy based on RDF generation cannot be adopted, the biological pre-treatment in areas with high selective collection of MSW should be avoided, according to Figure 2. Some comparative considerations are reported in Table 2.

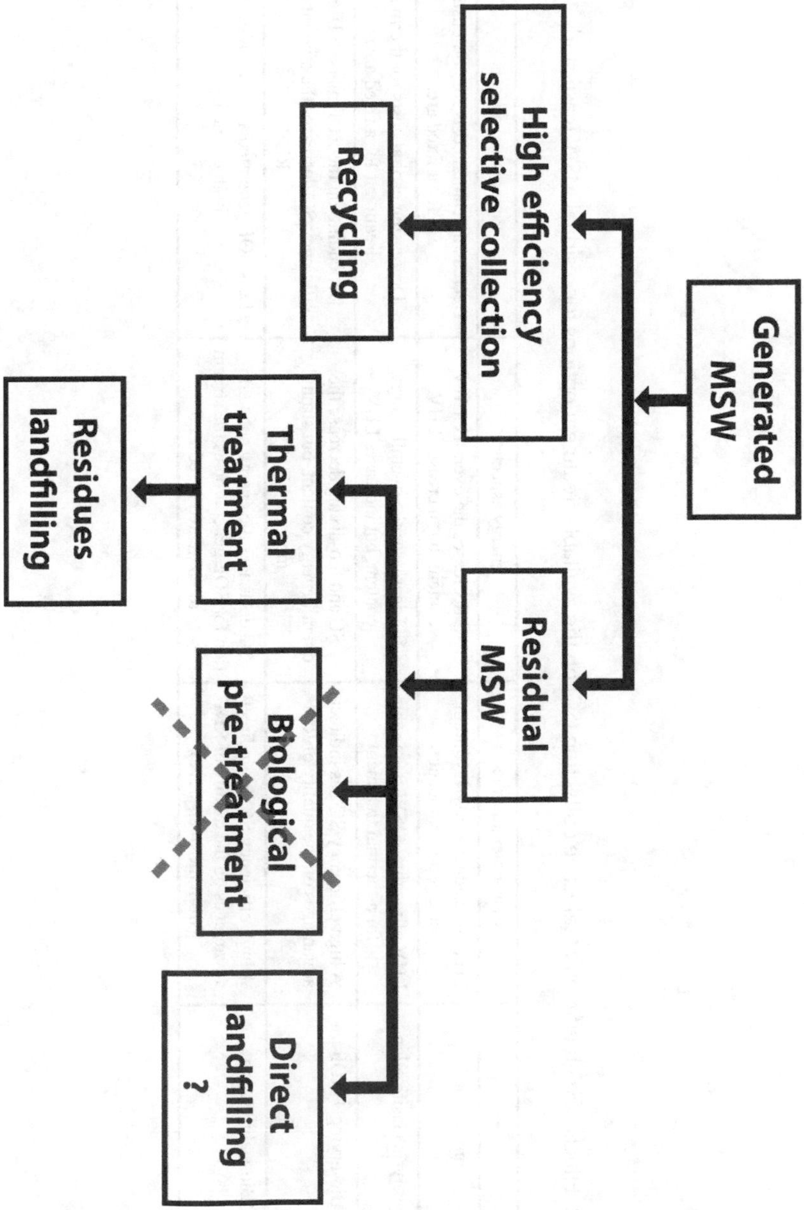

Figure 2. Effects of high efficiency selective collection.

Table 2. Effects of high efficiency selective collection (SC) on the available treatment options for residual MSW (case-study 1).

	Environmental aspects	energy aspects	Economical aspects
Combustion	LHV of residual MSW favourable for a good combustion	SC decreases the overall energy content, but increases LHV	Decreasing capital costs for combustion for a fixed area
Pyrolysis/Gasification	LHV of residual MSW favourable for a thermal treatment	SC decreases the overall energy content, but increases LHV	Decreasing capital costs for thermal treatment for a fixed area
Bio-drying + RDF	A high efficiency SC has indirect advantages on the quality of RDF	SC and bio-drying decrease the overall energy content, but strongly increase LHV	Depending on the RDF market. High efficiency SC decreases the amount of RDF.
Bio-stabilization	A high efficiency SC has indirect advantages on the quality of RDF (from the over-sieve)	SC decreases the energy consumption for SOF generation as the stream to be treated is reduced	Less SOF generated as result of SC at high efficiency

1.5 CASE STUDY 2 – ROLE OF CO_2

1.5.1 SCENARIO OF COMPARISON

This case-study refers to a comparison between waste-to-energy plant and a modern landfill taking into account the role of Carbon sink that could have a landfill where the residual MSW is affected by the high efficiency of selective collection. The characteristics of the residual MSW are the same of the first case-study. In particular, the content of carbon in the residual waste can be assumed as 25% of the total, with 15% of non-fossil origin and 10% of fossil origin. All the carbon of non-fossil origin is assumed to be converted into biogas, with the following content: 60% CH_4 and 40% CO_2. The average collection of biogas is assumed as 55% and it is converted into electricity with an efficiency of 40%. The impact of fugitive emission of CH_4 is assumed 23:1 compared to fossil CO_2. Concerning waste-to-energy plant, it has been assumed a combustion in a small plant, with a net electrical conversion equal to 20%. An additional assumption is that the generated electricity allows a saving of 600 $gCO_2\ kWh^{-1}el$ from the hypothetical national mix of fuels in the thermal power plants. Data can vary from country to country.

1.5.2 RESULTS AND DISCUSSION

Taking into account the LHV of residual MSW and the above parameters concerning the waste-to-energy scenario, the presence of biomass-like materials in the residual waste allows saving 67 gCO_2 kg-1.

Landfilling would seem to act as a Carbon sink option if we consider that a low putrescibility of waste could decrease the fugitive emissions of methane and plastics are inert when landfilled. On the contrary, looking at the characteristics of the residual MSW of the present case-study the putrescibility of the waste is not very low as demonstrated from the 19% of organic fraction content. In this scenario the CO_2 saving from electricity generation from methane is significant thanks to high efficiency of conversion obtainable with engines. The big problem of this strategy is related to the fugitive emissions of methane: the role of this stream is significant as this gas is highly impacting in terms of greenhouse effect. The amount of equivalent CO_2 related to methane emissions is widely higher than the above mentioned saving. The result is unfavorable for landfilling. This scenario could change if the selective collection of the organic fraction reaches higher values (in order to have a residual MSW poor of putrescible materials). Anyway it must be pointed out that the efficiency of collection of putrescible materials for the mentioned case is 85%, thus better results are not easy to be reached.

1.6 CONCLUSIONS

The results point out that an optimized selective collection decreases the list of available options for a correct management of MSW: the optimization of the source separation of the organic fraction makes it unviable the adoption of biological treatments for the residual MSW (as these processes need putrescible matter as engine of the system). In this scenario the role of the biological treatments should be limited to the processing of organic fraction selectively collected. To this concern the role of biological treatments is significant as the streams of putrescible materials selected at the source are increasing.

Additionally the characteristics of the residual MSW are interesting for an energy recovery, whilst the role of a direct landfilling is controversial. According to an European Union principle, landfilling of waste with high energy content (as residual MSW could be) is not allowed in spite of the potential role of carbon sink. Anyway the European Union criteria seem to be suitable even for high efficiency of selective collection of organic fractions.

REFERENCES

1. Ghezzi U., (2008). Trattamenti termici non convenzionali, Proceedings of 63° Corso di Aggiornamento in Ingegneria Sanitaria – Ambientale.
2. Rada E. C. (2005a). Municipal Solid Waste bio-drying before energy generation, PhD Thesis, University of Trento & Politehnica University of Bucharest

3. Rada E. C., Istrate I. A., Ragazzi M. (2009b). Trends in the management of the Residual Municipal Solid Waste, Environmental Technology, Vol. 30, No. 7, 651–661.
4. Rada E. C., Ragazzi M., Panaitescu V., Apostol T (2006). MSW bio-drying and bio-stabilization : an experimental comparison, Proceedings of ISWA 2005, International Conference: Towards integrated urban solid waste management system, CD version.
5. Rada E. C., Ragazzi M., Panaitescu V., Apostol T., (2005b). Bio-drying or bio-stabilization process ?, Stintific Buletin, seria C: Electrical Engineering, 67, 4, 51-60.
6. Schione G., (2008). Il riciclo e la sostenibilità ambientale dell'alluminio, Proceedings of 63° Corso di Aggiornamento in Ingegneria Sanitaria – Ambientale.

CHAPTER 2

Accounting for the Ecological Footprint of Materials in Consumer Goods at the Urban Scale

MEIDAD KISSINGER, CORNELIA SUSSMAN,
JENNIE MOORE, AND WILLIAM E. REES

2.1 INTRODUCTION

Residents of urban centers in high-income countries consume vast resources and generate immense volumes of waste to maintain their lifestyles [1,2]. Their levels of resource use and waste production make urban centers major contributors to global ecological change: increasing atmospheric carbon, global depletion of renewable and non-renewable resources, and ecological systems overload [3,4,5,6,7]. Many local governments are engaging in efforts to monitor and reduce these impacts, for example, using the ICLEI-Local Governments for Sustainability GHG Emissions Analysis Protocol [8] to measure their contributions to global greenhouse gas (GHG) emissions. While indeed such measurement is critical [9], we suggest that a comprehensive account of urban demands on global renewable resources and ecosystems is equally warranted. Urban ecological footprint analysis (EFA) aims to account for the full scope of energy and materials appropriated by a city's residents, businesses and operations; its biophysical inputs (i.e., various types of biologically productive land); and the carbon dioxide emissions produced as wastes. Indeed several cities have been engaged in measuring the size of their ecological footprints and using the ecological footprint as a policy communication tool: stressing the need for dramatic reform of urban based consumption and waste production toward ecologically sustainable levels. However, data requirements, including limited city-scale data, have prevented EFA from being fully employed as a local planning and monitoring tool [10,11].

In this paper we focus on one critical part of a city's ecological footprint, the material consumption component (i.e., materials consumed through purchase of products by city residents, local businesses and city operations). This component has been calculated to make up between 8% and 20% of some urban ecological footprints (e.g., [11,12,13,14,15]). Other major components of any urban footprint include: food consumption, use of electricity and natural gas, transportation, and water. The 'component, solid waste life cycle assessment approach' is an EFA approach that overcomes some data limitations by using data many cities regularly collect: municipal solid waste data which serves as a proxy for material consumption. The logic behind the approach is that most materials in consumer goods end up in the waste stream (either for disposal or recycling), some in a matter of minutes after consumption, others after a few years. Therefore, municipal waste reflects the quantities and types of materials consumed by a local population over a one-year period [12,16,17,18].The approach can only be used by cities with access to municipal solid waste data. Many cities in industrialized countries manage and report on municipal solid waste while cities in industrializing countries often do not. Our set of EF values for material types could still be useful for cities to identify the relative impacts of different materials.

In the component solid waste LCA approach, once material consumption types and quantities have been identified from the solid waste stream data, the ecological footprint for each material is calculated with energy use and CO_2 emissions data from process LCA studies.

Use of the approach could be made easier for local governments and planners with access to ecological footprint conversion factors (or EF values) for individual materials. The initial calculation would be a measure of biologically productive land required (in global hectares) required to produce one unit of the material. For a city to determine the size of the ecological footprint of its consumption of material x, per unit (gha) would be multiplied by the total units of consumption for one year.

To this end, we conducted a thorough review of LCA studies to develop a set of EF values for twelve materials commonly found in urban solid waste streams. We were able to generate a range of values for each material that reflects various production modes, production locations and study methods. These values should be seen as provisional given the small number of original process LCA studies available (Our review yielded 69 studies for 12 materials). Our set of EF values highlights the relative sizes of ecological footprints among materials and indicates that cities monitoring and planning for reductions in material consumption should be aware of both the quantity and ecological impact per unit of individual materials.

2.1.1. THE ECOLOGICAL FOOTPRINTS OF CITIES

From a biophysical perspective, cities are wholly dependent upon productive/assimilative land and waterscapes. High-income cities typically have eco-footprints several hundred times larger than their political or geographic footprints. As major consumers of resources and generators of waste, urban populations in these high-income cities are mainly responsible for humanity's current state of ecological overshoot [1,2]. Detailing the urban footprint is essential for sustainability planning. It helps to identify the major points of leverage for policies to reduce urban throughput of resources and production of wastes. It also helps to develop urban systems that contribute to the integrity of the ecosystems upon which cities depend.

EFA quantifies the biophysical 'load' that any specified population (or industrial process) imposes on its supportive ecosystems. The method uses data on energy and material consumption, waste generation (measured as carbon dioxide) and ecosystem productivity to estimate the total ecosystem area (in terms of global average hectares [gha]) required on a continuous basis to produce the resources consumed by the study population, and to assimilate its carbon emissions [19,20,21]. EFA can thus explicitly connect people's consumption, product by product, to several types of appropriated ecosystem areas, and reveal the gap (positive or negative) between local demand and global supply of bio-capacity.

2.1.2 ECOLOGICAL FOOTPRINT ANALYSIS AT THE CITY SCALE

To date, two main approaches have been developed to calculate ecological footprints at the sub-national scale: (i) the compound, (ii) the component. The compound approach uses national per capita ecological footprint data, and scales it to the city as much as possible, to reflect, for example, local energy sources. The advantage of this approach is that reliable national level data is more often available than city scale data. However, a footprint based on national level data is limited in its ability to reflect the impacts of local policy and action [10]. Within the component method, the dominant approach is to use economy-wide input-output analysis. We refer to this approach henceforth as the input-output sub-national approach. The input-output sub-national approach uses both local and national level data. It takes data on local expenditures (measured in units of national currency such as dollars) for some consumption items like food and materials, and relates them to carbon emissions in an extension of conventional monetary input-output analysis. This data is then combined with national ecological footprint assessment data. An alternative approach is to collect data directly from local government sources without relying on dollars as a proxy for consumption. We call this the direct approach. Because it relies on locally generated energy and materials flow data, this approach can reflect local changes in specific resource consumption and waste generation. For local policy makers and planners, this distinction is critical. They must be able to craft policies and actions that target local conditions, and to monitor and assess the effectiveness of these policies and actions over time. These requirements are not effectively met by EFA approaches based on national level data that are subsequently scaled to the city nor by input-output analysis that uses dollars as a proxy for actual consumption [13,16,22,23]. The direct component approach aims to use mostly local data. Local data are collected on specific components of the community/

city under study, for example, transportation, buildings, food, consumables and waste. It determines the ecological footprint of each and sums them to generate an overall urban ecological footprint [16,17,24]. Because the footprint is based on local data, the impact of local policy initiatives or actions can be measured over time with successive ecological footprint analyses. Cities often have access to or collect their own data on transportation use (e.g., average vehicle kilometers travelled by urban residents), and buildings (e.g., building types, their energy use and efficiency), which are required for the component analysis. More challenging to come by are data on personal consumption including food and material goods that are not derived from monetary expenditure statistics. This is where the 'component, solid waste LCA approach' is useful. It can be used to identify types and estimate quantities of materials consumed within a city over a specified period of time.

Cities that use the direct component approach will have information to communicate to urban residents and businesses about opportunities for reducing their footprints through wise consumption choices and conservation habits. Changes in consumption are critical to reforming cities toward biophysical sustainability. Material footprint information can be used to develop policy targeted toward strategic supply chain and product stewardship initiatives that avoid high-impact materials. It can support the development or expansion of local recycling and re-use programs, and encourage local businesses and institutions to identify ways their operations can reduce use of high impact materials through use of labeling and promotion of low-impact products. It can also identify when levels of material consumption exceed global ecological carrying capacity and inform how demand management efforts can curb escalating materials throughput in the urban metabolism.

2.1.3 THE COMPONENT SOLID WASTE APPROACH

The 'component, solid waste LCA approach' to EFA was developed by Simmons et al. [17,18], and has been implemented in several studies (e.g., [11,12,14,16,25]). While cities do not commonly monitor or document their residents' material consumption, many manage and monitor solid waste. Most materials that make up consumer goods will eventually enter the solid waste stream (for disposal or recycling), therefore over the course of a one year period, municipal waste reflects the quantities and types of materials consumed by a local population However, these data likely underestimate total amounts of materials consumed since all materials used in the manufacturing process are not reflected in the solid waste stream. Therefore, the approach incorporates lifecycle assessment data to estimate the amount of energy and materials used in the supply chain to manufacture the materials found in the waste stream. Although a small amount of waste may be disposed without use of municipal solid waste facilities, municipal solid waste can reasonably be expected to represent the majority of waste generated within the city by proportion and type.

Cities that monitor and document commercial and household waste composition data generally use the following major categories: metal; glass; plastics; paper; organics; textiles; rubber; and hazardous wastes. Many cities use more detailed categories. For example, paper is broken down into paper, newsprint, and cardboard. Plastics are identified by type (PET; HDPE; PVC and LDPE) and by use such as plastic (film) bags and plastic bottles (e.g., [26,27,28,29,30]). One consumer item that appears in the solid waste stream in high quantities and is commonly reported as a separate item is diapers (nappies). In Canberra, Australia in 2009, diapers made up 4.9%, by weight, of the residential waste stream [30]. A similar figure has been identified in several Israeli cities and towns [31].

Once material consumption types have been determined from the municipal solid waste stream, the ecological footprint for each is estimated. The calculations for these materials incorporate energy use and carbon emissions data from LCAs of the materials. Process LCA quantifies the diverse material and energy inputs required for the production of particular materials and products as well as various wastes including greenhouse gases (GHG), of which sometimes carbon dioxide (CO_2) emissions are uniquely distinguished [32,33,34]. This stage of the approach can be time consuming for local governments because it requires a review of LCA studies and the footprint calculations. Ecological footprinting could be made easier for cities through development of an open source, international database of ecological footprint values for materials. City staff would use their own solid waste data to identify quantities of materials consumed by weight over a given period of time, and look up the ecological footprint value for each material in the database. All data sources and calculations could be transparent and accessible for users to access as well.

A set of ecological footprints of individual materials and processes has been published by Huijbregts et al. [35]. Values in that study are derived from the Swiss ecoinvent database. In absence of an open source, international database, that source ecological footprint values could be a starting point for cities. However, it [35] relies heavily on European process LCA data. A dataset of ecological footprint values for materials should reflect production practices around the world including China, a major manufactured goods exporter. Therefore, in our review of process based LCA studies, we included some of the studies that support that dataset [35], but also searched explicitly for studies from Asia, North America, South and Latin America and Australia. In calculating the ecological footprint, Huijbregts et al. [35] separate the energy component into carbon sequestration lands and demand on ecosystems from nuclear energy sources; this separation is no longer standard [36].

Another approach to measuring the EF of material consumption has been proposed by Herva et al. [37]. This approach aims to account for hazardous and toxic wastes that have been combusted with thermal plasma technology. Herva et al. suggest the EF approach is more suitable for industry than for municipal policy makers.

The solid waste LCA approach has limitations that arise from reliance on process LCA studies. While some standardization of method exists, LCA specialists regularly highlight the many assumptions required for LCA, and the limitations to comparison of results between studies [38]. As more material LCA studies are published under increasingly standardized conditions, confidence in results of the solid waste LCA based approach will increase.

2.2 METHODS

To identify materials commonly consumed in cities, and for which we would determine ecological footprint conversions factors, we reviewed reports on solid waste composition for 10 cities in the UK, the US, Canada, Israel, and Australia (Vancouver (CAN); Edmonton (CAN); Sidney (AUS); Melbourne (AUS); London (UK); York (UK); Edinburgh (UK); Haifa & Ashdod (ISR); Denver (USA); Seattle (USA). Six material types appeared most often: (1) paper (2) plastics (3) diapers (4) glass (5) metals and (6) textiles. The materials were represented as a percentage of the solid waste streams, by weight, in the following order: paper made up the largest percentage in all cities, ranging from 11.5% to 27.8%; followed by various types of plastics (ranging from 12.1% to 16.9%). Glass, metals, and textiles were the smallest. Five cities reported on diapers (Vancouver, Canberra, Melbourne, Edinburgh; Haifa), and in each this material was the third largest by weight ranging from 3.1% in Melbourne to 5.26% in Vancouver.

From the material types reported by the cities we selected twelve materials: (1) print paper, (2) newsprint, (3) cardboard, (4) PET, (5) HDPE, (6) PVC, (7) PS, (8) glass (9) diapers, (10) cotton textile, (11) aluminum, and (12) steel.

We conducted an extensive review of LCA studies for each material, examining academic literature, and commercial and industrial reports. Each LCA study sets its own boundaries and scale. In order to present comparable footprint and emissions values we made an effort to include studies that used similar parameters, assumptions, and scales. Overall we made an effort to cover cradle to gate data. This means data associated with the manufacturing process from materials extraction to finished product that leaves the factory gate. In the case of plastics most of our values are for plastic polymers owing to lack of available LCA data on finished products. The review yielded 69 relevant studies and included data for European, North American, Asian, and Australian production locations among others. For the complete list of studies see Appendix I. From each LCA study, for each material, we extracted the data on energy sources and CO_2 emissions for our calculations of ecological footprint values. While most studied materials are from non-renewable resources (e.g., plastic, metals) and therefore their EF include only energy land (i.e., the area of land for sequestering the carbon dioxide emitted), for materials from renewable resources (e.g., paper, cotton) the study also calculated the crop and forest land required.

2.2.1 FROM LCA TO FOOTPRINT

The ecological footprint comprises several types of land including cropland, forest land and energy land [39,40,41]. The LCA literature review provided us with data on the energy and CO_2 emissions associated with each material's life cycle. These data were used for the conversion to ecological footprint. The energy land was calculated for all materials; in the case of materials made from renewable sources, e.g., paper and textiles, the area of forest land (paper) and cropland (textiles) was also calculated.

Energy land: For calculating the energy land footprint (i.e., the land required to sequester carbon emissions) we used data on the CO_2 per tonne of material/product as reported in the LCA studies. We then used a conversion factor of 0.27 hectares of world forest, which is the average area of forest land required to sequester 1 tonne of CO2 [15,42,43]. From that, one-third was deducted to reflect the emissions absorbed by the oceans [43]. Finally, an equivalence factor of 1.24 for energy land was applied (Following [15,42]).

Crop land: The cropland footprint has been assessed for cotton textiles. A simplified equation displayed below explains the cropland footprint calculation procedure:

$$EF_{(gha)} = (U/GY) \times EQF \quad (1)$$

where the footprint was calculated by dividing the unit (U) of consumption (1 tonne) by the annual average global yield (GY) [44] and then multiplied by an equivalence factor of 2.39 for cropland [42].

Forest land: To calculate the EF of paper products we integrated several sources. Data on energy inputs and carbon emissions were taken from the LCA papers. The carbon dioxide was then converted to energy land as described above. We then used the UNECE/FAO [45] forest products conversion factors to obtain an average figure for the amount of wood required to produce different types of paper (m3 of wood/metric tonne of paper). The next step was to use Global Footprint Network [42] data to convert the equivalent wood weight to the required forest land. The conversion figure used is 1.81 m^3 of wood per hectare of forest with global average productive yield. Finally, that area of forest land was multiplied by an equivalence factor of 1.24 [42].

2.2.2 LIMITATIONS

Our ecological footprint dataset is limited by the small number of original LCA studies available for each material from international sources. In particular, we found few published LCA studies on materials produced in China. This means that production systems in one of the world's largest exporting nations are under-represented in our set of ecological footprint values. We also found an underwhelming number of studies on some materials, like diapers, that make up a significant component (by weight) of many urban waste streams.

EFA only includes carbon dioxide emissions which, unlike other GHGs, are continually being taken up by natural systems such as forests and oceans. In our review of LCA studies we found that several reported only carbon dioxide equivalent (CO_2e) associated with a material's production. CO_2e includes GHG emissions such as methane and nitrous oxide. Because we had no reliable method for disaggregating the CO_2 from CO_2e, we could not use these studies. We do include the studies and their CO_2e data in the Appendix because they represent the findings of our LCA review, and to reveal the scope of available data.

Because our data set is established from published LCA studies and reports, any inherent issues of data quality, system parameters, or calculations, are repeated in our review. We attempted to overcome this limitation by including a range and breadth of LCA studies, and also by presenting the range of EF values rather an average for each material. We restricted our LCA boundaries from cradle to gate. We did not include components such as shipping distances to the city or the footprint of the retailer (this would be part of the commercial building footprint component). Therefore, the size of the footprints is likely underestimated.

2.3 RESULTS AND DISCUSSION

Table 1 summarizes the ecological footprint values (measured in gha/tonne) that we calculated for the most common materials found in municipal waste streams. The minimum, maximum and mean emissions for each material are shown together with the standard deviation.

Consolidating the data in a table enables identification of high and low impact material. For example, textiles have the highest ecological footprint per unit of material. This is because of the large area of agricultural land and energy inputs required to grow cotton, as well as high amounts of energy used in the material's production. Paper, cardboard and newsprint also have relatively high ecological footprints due to the land area required to grow trees. The actual footprint values for plastics are likely higher than the values reported here because over half of the data sources available for plastics are for plastic polymers rather than finished products.

Urban policy makers, planners, businesses and residents will want to link these data with their local solid waste data that identifies how much of each material is regularly consumed. Although textiles and aluminum have high ecological footprints, they make up smaller proportions of most urban waste streams than paper products and diapers. Consumption of paper products and diapers might be more important targets for reduction at the urban scale. This level and detail of information is the kind required by policy makers and planners who wish to develop targeted local policies and by local residents, businesses and organizations seeking to reduce their urban ecological footprints.

In our review of LCA studies we found variations in LCA measures from different parts of the world. While these variations can be explained by different LCA methods and data availability, they may also reflect variations in production methods and energy sources used in different parts of the world (e.g., coal based electricity vs. hydroelectric sources). For example, the EF of a ton of aluminum produced in China or in Australia is at least 45% larger than one tonne of aluminum produced in Europe. Publication of more LCA studies is necessary for true comparisons. In a previously published paper [9], we presented the GHG emissions calculated for the same set of materials, using data from the same

Table 1. Ecological footprint of materials.

	N	Min	Max	Average	Standard deviation
Sub Category		gha/tonne	gha/tonne	gha/tonne	gha/tonne
Glass	5	0.18	0.40	0.24	0.09
HDPE	4	0.13	0.35	0.20	0.10
PVC	4	0.31	0.53	0.41	0.07
PET	5	0.24	0.57	0.48	0.11
Polystyrene	5	0.26	1.04	0.66	0.25
Steel	10	0.38	0.90	0.61	0.16
Diapers	2	0.94	1.34	1.14	0.28
Newsprint	6	2.15	2.34	2.23	0.08
Cardboard	3	2.71	2.79	2.76	0.04
Printing paper	12	2.62	3.23	2.82	0.17
Aluminum	5	1.77	4.06	2.42	0.75
Cotton fabric	8	8.35	12.21	10.20	1.59

LCA sources. We believe it is important for cities to have information on both GHG emissions and ecological footprint values to determine (1) sustainable levels of urban material consumption and (2) to inform policies and programs that promote rapid change toward those levels. We compare GHG emissions values with average ecological footprints for the twelve materials in Table 2, arranged in ascending order of size/emissions per tonne.

Table 2 highlights the differences in environmental impact each measurement tool is designed to assess: GHG emissions inventories account for the GHG emissions related to a material's production, while the ecological footprint accounts for the area of biologically productive land required to produce a material and to assimilate associated production wastes, in perpetuity. The difference is apparent

Table 2. Average ecological footprints and greenhouse gas (GHG) emissions of materials.

Average GHG Emissions		Average Ecological Footprint	
Sub Category	**CO_2e/ tonne**	**Sub Category**	**gha/tonne**
Cardboard	890	Glass	0.24
Glass	990	HDPE	0.20
HDPE	1,015	PVC	0.41
Newsprint	1,120	PET	0.48
Printing paper	1,290	Steel	0.61
PVC	1,920	Polystyrene	0.66
PET	2,240	Diapers	1.14
Steel	2,530	Newsprint	2.23
PS	2,970	Cardboard	2.76
Diapers	3,580	Printing paper	2.82
Aluminum	10,840	Aluminum	2.42
Cotton fabric	21,500	Cotton fabric	10.20

in a comparison of materials. Cotton textile is a high impact material in terms of both ecological footprint and GHG emissions. Cotton textiles require land area on which to grow cotton crops, and textile production is energy intensive. Aluminum has an ecological footprint approximately equal in size to that of newsprint or cardboard. Its GHG emissions, however, are almost ten times higher. In ecological footprint accounting, the tremendous energy requirements of aluminum are registered as ‘energy land’ (bio-productive land area to sequester carbon dioxide) and make up most of the aluminum footprint. However, paper production impacts more heavily on use and over-use of the earth’s bio-productive land area. Paper products are less energy intensive to produce so the ‘energy’ land component of their ecological footprints is smaller, but the size of their ecological footprints is increased by their requirement for 'forest land' on which to grow trees. Similar to aluminum, the plastics: PET and PS have over twice the GHG emissions of cardboard, newsprint and printing paper. Therefore, production of plastics (per ton) makes a greater contribution to atmospheric carbon levels, yet their ecological footprints are approximately one quarter the size of any paper products.

In our review of the literature, we found few academic studies whose sole aim was to carry out a product or material LCA. More often, the study purpose was to investigate potential improvements in production methods or management systems, for example, comparing different textile tailoring scenarios [46], or end of life options for plastic bottles [47]. In these studies boundaries often excluded production of the material itself. Other studies used existing material LCA data, such as those found in ecoinvent. Studies on production processes outside of Europe frequently included ecoinvent LCA data which suggests challenges in obtaining local process data. Our review of Chinese language journals yielded five appropriate studies. In total we found 69 LCA studies that provided data necessary for use in our ecological footprint conversion calculations. The lack of studies aimed directly at assessing life cycle production impacts suggests that promotion of their value for sustainability research and action could be heightened. It also speaks to the challenges posed by high data requirements and the lack of reliable, accessible data, including the barrier of proprietary data related to some industrial processes. We found an increasing number of material LCAs conducted or commissioned by commercial and industrial associations such as the World Aluminum Association and the European Container Glass Federation. Individual companies are also publishing information on the carbon emissions of their products. For these products, the actual LCA studies are not commonly available but in some cases, information such as the study author, reviewer, or LCA protocol is provided. Perhaps more industry-based studies will be conducted as carbon taxes and cap and trade systems are expanded. Consumer pressure for more ecologically benign products may also encourage more reporting.

2.4 CONCLUSIONS

With data from the 69 LCA studies, we were able to develop a range of ecological footprint values for each material, reflecting production data from several countries. Our set of EF values incorporates the best currently available LCA data. Given the small number of studies for each material, and from each production location, and the differences in study methods, our values should be viewed as provisional. A greater number of standardized LCA studies using consistent process boundaries will increase accuracy in measuring ecological footprints associated with the twelve materials. Cities accounting for the ecological footprints of their material consumption could follow the EFA convention of selecting the lowest reported value for each material, and up-dating values as LCA data become available. The relative size of ecological footprints (per production unit) among the materials is likely a fair representation and offers cities a general ranking of materials that may be useful in policy and program development.

We believe that the ecological footprint with its comprehensive representation of ecological loads that extend beyond carbon emissions should become as mainstream a local policy and planning tool as GHG emissions inventories have become over the last decade. While climate change is a critical issue for cities to tackle through GHG emissions reduction policy and action, increasing global resource depletion and related ecosystem impacts pose equally imminent and dramatic risks [2,47] EFA reveals that the global human population is using earth's resources more quickly than they can be replenished. In fact, it would take the earth 1.5 years to regenerate the resources consumed and assimilate the wastes produced in 2012 [42]. As discussed, EFA at the urban scale must overcome challenges related to local data availability. The component solid waste LCA approach can overcome this limitation for the material consumption component, for those cities that monitor municipal solid waste. Local government use of the solid waste LCA approach helps to clearly identify the ecological impacts associated with the waste they manage on behalf of their residents. This direct connection can be used to communicate to citizens about stewardship, recycling and ecologically responsible consumption choices that contribute to urban sustainability.

REFERENCES

1. WWF (World Wide Fund for Nature), Living Planet Report; World Wide Fund for Nature: Gland, Switzerland, 2010.
2. WWF (World Wide Fund for Nature), Living Planet Report; World Wide Fund for Nature: Gland, Switzerland, 2012.
3. Alberti, M. Measuring urban sustainability. Environ. Impact Assess. Rev. 1996, 16, 381–424.
4. Rees, W.E. Is sustainable city an oxymoron? Local Environ. 1997, 2, 303–310.
5. Rees, W.E. Getting serious about urban sustainability: Eco-footprints and the vulnerability of 21st century cities. In Sustainability Science: The Emerging Paradigm and the Urban Environment; Bunting, T., Filion, P., Walker, R., Eds.; Springer: Toronto, Canada, 2010. Chapter 5.
6. Newman, P. The Environmental Impacts of Cities. Environ Urban. 2006, 18, 275–295.
7. Grimm, B.N.; Faeth, H.S.; Golubiewski, E.N.; Redman, L.C.; Wu, J.; Bai, X.; Briggs, M.J. Global change and the ecology of cities. Science 2008, 319, 756–760.
8. Local Governments for Sustainability (ICLEI). International Local Government GHG Emissions Analysis Protocol (IEAP). Version 1.0.. 2009. Available online: http://carbonn.org/fileadmin/user_upload/carbonn/Standards/IEAP_October2010_color.pdf (accessed on 26 April 2013).
9. Kissinger, M.; Sussman, C.; Moore, J.; Rees, W.E. Accounting for GHG of materials at the urban scale—Relating existing process life cycle assessment studies to urban material and waste composition. LCE 2013. in press.
10. Wilson, J.; Grant, J. Calculating ecological footprints at the municipal level: What is a reasonable approach for Canada? Local Environ. 2009, 14, 963–979.
11. Moore, J.; Kissinger, M.; Rees, W.E. An urban metabolism assessment and ecological footprint of metro vancouver. J. Environ. Manage. 2013, 124, 51–61.
12. Barrett, J.; Vallack, H.; Jones, A.; Haq, G. A Material Flow Analysis and Ecological Footprint of York; Technical Report. Stockholm Environmental Institute: Stockholm, Sweden, 2002.
13. Aall, C.; Norland, I. The Use of the Ecological Footprint in Local Politics and Administration: Results and implications from Norway. Local Environ. 2005, 10, 159–172.
14. Kissinger, M.; Haim, A. Urban Hinterlands: The case of an Israeli town ecological footprint. Environ. Dev. Sustain. 2008, 10, 391–405.
15. Scotti, M.; Bondavalli, C.; Bondini, A. Ecological footprint as a tool for local sustainability: The municipality of Piacenza (Italy) as a case study. Environ. Impact Assess. 2009, 29, 39–50.

16. Chambers, N.; Simmons, C.; Wackernagel, M. Sharing Natures Interest; Earthscan: London, UK, 2000.
17. Simmons, C.; Chambers, N. Footprinting UK households: How big is your ecological garden? Local Environ. 1998, 3, 355–362.
18. Simmons, C.; Lewis, K.; Barrett, J. Two feet—two approaches: A component-based model of ecological footprinting. Ecol. Econ. 2000, 32, 375–380.
19. Rees, W.E. Ecological footprints and appropriated carrying capacity: What urban economics leaves out. Environ. Urban. 1992, 4, 121–130.
20. Rees, W.E.; Wackernagel, M. Ecological footprints and appropriated carrying capacity: measuring the natural capital requirements of human economy. In Investing in Natural Capital: The Ecological Economics Approach to sustainability; Janson, A.M., Hammer, M., Folke, C., Costanza, R., Eds.; Island press: Washington, DC, USA, 1994; pp. 362–390.
21. Wackernagel, M.; Rees, W.E. Our Ecological Footprint—Reducing Human Impact on the Earth; New Society Publishers: Gabriola, BC, Canada, 1996.
22. Curry, R.; Maguire, C.; Simmons, C.; Lewis, K. The use of material flow analysis and the ecological footprint in regional policy-making: Application and insights from Northern Ireland. Local Environ. 2011, 16, 165–179.
23. Wiedmann, T.; Minx, J.; Barrett, J.; Wackernagel, M. Allocating ecological footprints to final consumption categories with input-output analysis. Ecol Econ. 2006, 56, 428–448.
24. Barrett, J. Component ecological footprint: Developing sustainable scenarios. Impact Assess. Proj. Apprais. 2001, 19, 107–118.
25. Gottlieb, D.; Kissinger, M.; Haim, A.; Vigoda, E. Implementing the ecological footprint at the institute level. Ecol Indic. 2011, 18, 91–97.
26. Cascadia Consulting Group. Seattle Public Utilities 2010 residential waste stream composition study final report. 2011. Available online: http://www.seattle.gov/util/Documents/Reports/SolidWasteReports/CompositionStudies/index.htm/ (accessed on 17 March 2012).
27. City of Edinburgh. The City of Edinburgh Council Edinburgh's waste and recycling strategy 010–2025; City of Edinburgh: Edinburgh, UK, 2010.
28. City of Edmonton. The Edmonton sustainability papers discussion paper 10—Sustainable waste management; City of Edmonton: Edmonton, Canada, 2010.
29. AEA Technology, Greater London Authority Waste Composition Scoping Study; AET/ENV/R/1826; AET Technology: Oxfordshire, UK, 2004.
30. APC, Kerbside Domestic Waste and Recycling Audit; APC Environmental Management: Sydney, Australia, 2009.
31. IMEP—The Israeli Ministry of Environmental Protection. National solid waste composition survey; The Israeli Ministry of Environmental Protection: Jerusalem, Israel, 2005.
32. Udo de Haes, H.A.; Jolliet, O.; Finnveden, G.; Hauschild, M.; Krewitt, W.; Muller-Wenk, R. Best available practices regarding impact categories and category indicators in life cycle impact assessment—Part I. Int. J. Life Cycle Ass. 1999, 4, 66–74.
33. Halberg, N.; Weidema, B. Life Cycle Assessment in the Agrifood Sector. 2004. Available online: http://www.lcafood.dk/lca conf/ (accessed on 12 June 2011).

34. Berg, S.; Lindholm, E. Energy use and environmental impacts of forest operations in Sweden. J. Clean Prod. 2005, 13, 33–42.
35. Huijbregts, M.; Hellweg, S.; Frischknecht, R.; Hungerbuhler, K.; Hendriks, A. Ecological footprint accounting in the life cycle assessment of products. Ecol. Econ. 2008, 64, 798–807.
36. Global Footprint Network, Ecological Footprint Standards 2009; Global Footprint Network: Oakland, CA, USA, 2009.
37. Herva, M.; Hernando, R.; Carrasco, E.; Roca, E. Development of a methodology to assess the footprint of wastes. J. Hazard Mater. 2010, 180, 264–273.
38. Boustead, I. Eco-profiles of the European Plastics Industry: Low Density Polyethylene; PlasticsEurope: Brussels, Belgium, 2005.
39. Galli, A.; Kitzes, J.; Wermer, P.; Wackernagel, M.; Niccolucci, V.; Tiezzi, E. An Exploration of the mathematics behind the ecological footprint. In Ecodynamics: The Prigogine Legacy; Brebbia, C., Ed.; Wit Press: Billerica, MA, USA, 2007; pp. 249–256.
40. Wackernagel, M.; Monfreda, C.; Schulz, N.; Erb, K.; Haberl, H.; Krausmann, F. Calculating national and global ecological footprint time series: resolving conceptual challenges. Land Use Policy. 2004, 21, 271–278.
41. Kitzes, J.; Peller, A.; Goldfinger, S.; Wackernagel, M. Current methods for calculating national ecological footprint accounts. Sci. Environ. Sust. Soc. 2007, 4, 1–9.
42. Global Footprint Network. Footprint Basics Overview. 2012. Available online: http://www.footprintnetwork.org/en/index.php/GFN/page/footprint_basics_overview/ (accessed on 22 August 2012).
43. Monfreda, C.; Wackernagel, M.; Deumling, D. Establishing national natural capital accounts based on detailed ecological footprint and biological capacity assessments. Land Use Policy 2004, 21, 231–246.
44. FAO (United Nations Food and Agriculture Organization). FAOSTAT: Production: Crops 2010. Available online: http://faostat.fao.org/site/567/DesktopDefault.aspx/ (accessed on 2 May 2012).
45. UNECE/FAO. Forest Product Conversion Factors for the UNECE Region; Timber section: Geneve, Switzerland, 2009.
46. Herva, M.; Franco, A.; Ferreiro, S.; Alvarez, A.; Roca, E. An approach for the application of the ecological footprint as environmental indicator in the textile sector. J. Hazard Mater. 2008, 156, 478–487.
47. Millennium Ecosystem Assessment. Ecosystems and Human Well-being: Synthesis; Island Press: Washington, DC, USA, 2005. Available online: http://www.maweb (accessed on 7 May 2012).

CHAPTER 3

Advances on Waste Valorization: New Horizons for a More Sustainable Society

RICK ARNEIL D. ARANCON, CAROL SZE KI LIN, KING MING CHAN, TSZ HIM KWAN, AND RAFAEL LUQUE

3.1 INTRODUCTION

Climate change, energy crisis, resource scarcity, and pollution are major issues humankind will be facing in future years. Sustainable development has become a priority for the world's policy makers since humanity's impact on the environment has been greatly accelerated in the past century with rapidly increasing population and the concomitant sharp decrease of ultimate natural resources. Finding alternatives and more sustainable ways to live, in general, is our duty to pass on to future generations, and one of these important messages relates to waste. Waste from different types (e.g., agricultural, food, industrial) is generated day by day in extensive quantities, generating a significant problem in its management and disposal. A widespread feeling of "environment in danger" has been present everywhere in our society in recent years, which, however, has not yet crystallized in a general concienciation of cutting waste production in our daily lives. Many methods could achieve sustainable development, methods that could not only improve waste management but could also lead to the production of industrially important chemicals, materials, and fuels, in essence, valuable end products from waste.

Waste valorization is the process of converting waste materials into more useful products including chemicals, materials, and fuels. Such concept has already existed for a long time, mostly related to waste management, but it has been brought back to our society with renewed interest due to the fast depletion of natural and primary resources, the increased waste generation and landfilling worldwide and the need for more sustainable and cost-efficient waste management protocols. Various valorization techniques are currently showing promise in meeting industrial demands. One among such promising waste valorization strategies is the application of flow chemical technology to process waste to valuable products. A recent review of Ruiz et al. [1] highlighted various advantages of continuous flow processes particularly for biomass and/or food waste valorization which included reaction control, ease of scale-up, efficient reaction cycles producing more yield, and no required catalyst separation. Although flow chemistry has been known to be used in industries for other processing methodologies, it still remains to be used in biomass/waste valorization – a limitation caused by the large energy needed to degrade highly stable biopolymers and recalcitrant compounds (e.g., lignin). The deconstruction of such biopolymers, most of the time, requires extreme conditions of pressure and temperature – conditions achieved by microwave heating, which is another green valorization technology. These requirements are not simple to satisfy and various techniques (e.g., microwave irradiation) need to be combined to satisfy the prerequisites for a successful transformation of waste. However, the main challenge for this combination is on the scale-up itself. As conceptualized by Glasnov et al. [2] microwave and flow chemistries maybe coupled by attaching back-pressure regulators to flow devices. This approach can revolutionize industrial valorization since it will synthesize products fast (due to microwave heating) on one continuous run (flow process). Although the approach presented is possible, the main challenge of temperature transfer from microwave to flow remains to be solved. A buildup of temperature gradient inside the instrument could lead to various instrument inefficiencies.

Another valorization strategy is related to the use of pyrolysis in the synthesis of fuels. This involves biomass heating at high temperatures in the absence of oxygen to produce decomposed products [3, 4]. Although pyrolysis is a rather old method for char generation, it has been recently utilized to produce usable smaller molecules from very stable biopolymers. This method has been particularly employed in the production of Bio-Oil (a liquid, of relatively low viscosity that is a complex mixture of short-chain aldehydes, ketones, and carboxylic acids). In a study by Heo et al. [5], several conditions for the fast pyrolysis of waste furniture sawdust were studied, and it was found that bio-oil yields do not necessarily increase with temperature. The optimized pyrolysis temperature was set at 450°C (57% bio-oil yield) using a fluidized bed reactor. The reason for the nonlinear dependence bio-oil yield/temperature is the possible decomposition of small molecules into simpler gases. This theory is supported by the increase in the amount of gaseous products found at increasing temperatures. A separate study by Cho et al. [6] employed fast pyrolysis under a fluidized bed reactor to recover BTEX compounds (benzene, toluene, ethylbenzene, and xylenes) from mixed plastics. The highest BTEX yield was obtained at 719°C. The pyrolysis of cotton stalks was also reported to produce second generation biofuels [7]. The study found that at much higher temperatures of pyrolysis the amounts of H_2 and CO collected increased, while CO_2 levels lowered. The decrease in CO_2 production could be due to the degradation of the gas at much higher temperatures producing CO and O_2. More recently, synergy between these first proposed technologies (microwave and pyrolysis) has been also reported to constitute a step forward toward more environmentally friendly low temperature pyrolysis protocols for bio-oil and syngas production [8]. Microwave-assisted pyrolysis of a range of waste feedstock can provide a tuneable and highly versatile option to syngas with tuneable H_2/CO ratios or bio-oil-derived biofuels via subsequent upgrading of the pyrolysis oil [8].

Aside from energy applications, pyrolysis can also be used to produce advanced materials including carbon nanotubes and graphene-like materials, which have a wide range of applications. These studies along with many others in literature illustrate the potential of pyrolysis to convert waste materials into valuable chemicals.

A third green method of valorization would be on the use of biological microorganisms to degrade complex wastes and produce fuel. The method is used by taking advantage of cellulose (or any biopolymer) degrading enzymes by microorganisms as demonstrated by Wulff et al. [9]. In their work, cellulase Xf818 was isolated from the plant pathogen *Xylella fastidiosa* (known to cause citrus variegated chlorosis in plants). The gene responsible for the enzyme was also probed and then later on expressed on *Escherichia coli*. Such enzyme was found to be mostly active in the hydrolysis of carboxymethyl celluloses, oat spelt xylans, and wood xylans.

Bioconversion has been under intensive research for the past years, and one of the most significant advances in the field relates to the possibility of a synthetic control of microorganisms' metabolic pathways to produce favorable metabolic processes, which will in turn increase the yield of products. A notable example is the use of a bioengineered *E. coli* to produce higher alcohols including isobutanol, 1-butanol, 2-methyl-1-butanol, 3-methyl-1-butanol and 2-phenylethanol from glucose [10]. The protocol was amenable for the conversion of 2-ketoacid intermediates (from amino acid biosynthesis) into alcohols by amplifying expression of 2-ketoacid decarboxylases and alcohol dehydrogenases.

To model the design for isobutanol, the gene ilvIHCD was over-expressed with the $P_L lacO_1$ promoter in a plasmid to amplify 2-ketoisovalerate biosynthesis. Other genes were tested such as alsS gene (from *Bacillus subtilis*) to further improve the alcohol yield while some genes responsible for by-product formation (*adhE*, *ldhA*, *frdAB*, *fnr*, *pta*) and pyruvate competition (*pflB*) were silenced. Overall, the isobutanol yield reached ~300 mmol/L (22 g/L) under microaerobic conditions.

The three presented strategies (microwave, pyrolysis, and bioengineering) represent some of the most important valorization methodologies. With the rapid advancement of these fields in waste valorization, it is expected that most industrial sustainability practices will have a different focus in various future scenarios.

Waste valorization is currently geared toward three sustainable paths: one would be on the production of fuel and energy to replace common fossil fuel sources and in parallel on the production of high-value platform chemicals as well as useful materials. Fossil-based fuels are clearly diminishing in supply and this has caused a global environmental concern due to rapidly rising emissions of fossil fuel by-products (both for processing and actual use). Because of this, waste valorization for energy and fuels are not only geared toward a sustainable fuel source but also toward a more benign fuel fit for an industrial up-scale. According to the Netherlands Environmental Assessment Agency, global CO_2 emissions reached an all-time high in 2011 at around 34 billion tonnes of greenhouse gases (GHGs) [3]. Close to 90% of these emissions derive from fossil fuel combustion. Other toxic gases such as volatile organic compounds, nitrogen oxides (precursors of toxic ozone) and particulates come together with GHGs. In a more than likely scenario of a minimum of 2.5% energy demand growth per year, it is necessary to substitute fossil fuels progressively with cleaner fuel sources. Biomass combustion for electricity and heat production was reported to be less costly, providing at the same time a larger CO_2-reduction potential [11]. Many studies also have shown convincing proof that the use of biomass for energy applications could be a highly interesting solution and cleaner technology for the future [12-14].

Another direction of waste valorization aims to produce high-value chemicals from residues including succinic acid (SA) [15], furfural and furans [16], phenolic compounds [17], and bioplastics [18]. These can be produced via chemical, chemo-enzymatic, and biotechnological approaches (e.g., solid state fermentation) but depending on the type of residue some compounds (e.g., essential oils, chemicals, etc.) can even be produced upon extraction and isolation [19]. The production of biomass-derived chemicals is a sustainable approach since it maximizes the use of resources and, at the same time, minimizes waste generation.

The major strength of biotechnology is its multidisciplinary nature and the broad range of scientific approaches that it encompasses. Among the broad range of technologies with the potential to reach the goal of sustainability, biotechnology could take an important place, especially in the fields of food production, renewable raw materials and energy, pollution prevention, and bioremediation. At present, the major application of biotechnology used in the environmental protection is to utilize microorganisms to control environmental contamination. Developing biotechnology could be a solution for these problems – this will also be given emphasis in this review.

Although waste valorization is an attractive approach for sustainability, on a large scale perspective, the purification, processing, and even the degradation of stable natural polymers (e.g., lignin) into simple usable chemicals still remain a significant challenge (Fig. 1).

In recent years, there have been increasing concerns in the disposal of food waste. The amount of food waste generated globally accounts for a staggering 1.3 billion tonnes per year. Apart from causing the loss of a potentially valuable food source or the regenerated resource, there are problems associated with the disposal of food waste into landfills. With this imminent waste management issue, food waste should be diverted from landfills to other processing facilities in the foreseeable future. In Hong Kong, there are 3600 tonnes of food waste generated (Table 1), 40% of which is made up of municipal solid waste (MSW). Fifty two percent (52%) of the MSW generated is dumped into landfills [20]. It is estimated that by 2018, all current landfill sites in Hong Kong will be exhausted.

Although the problem of food waste is commonly found over the world, the systems of food waste processing can only be formulated at the local community level with the consideration of the area-specific characteristics. These include regional characteristics and composition of waste, land availability, people's attitude and so forth. However, due to the lack of the local study concerning the suitability of food waste processing technologies for Hong Kong, this review is important to provide a few suggestions for the authorities to contemplate the adoption of a strategy on food waste disposal.

This contribution has been conceived to provide an overview on recent development of waste valorization strategies (with a particular emphasis on food waste) for the sustainable production of chemicals, materials, and fuels, highlighting key examples from recent research conducted by our groups. Reports on the development of green production strategies from waste and key insights into the recent legislation on management of wastes worldwide will also be discussed. The incorporation of these processes in future biorefineries for the production of value-added products and fuels will be an important contribution toward the world's highest priority target of sustainable development.

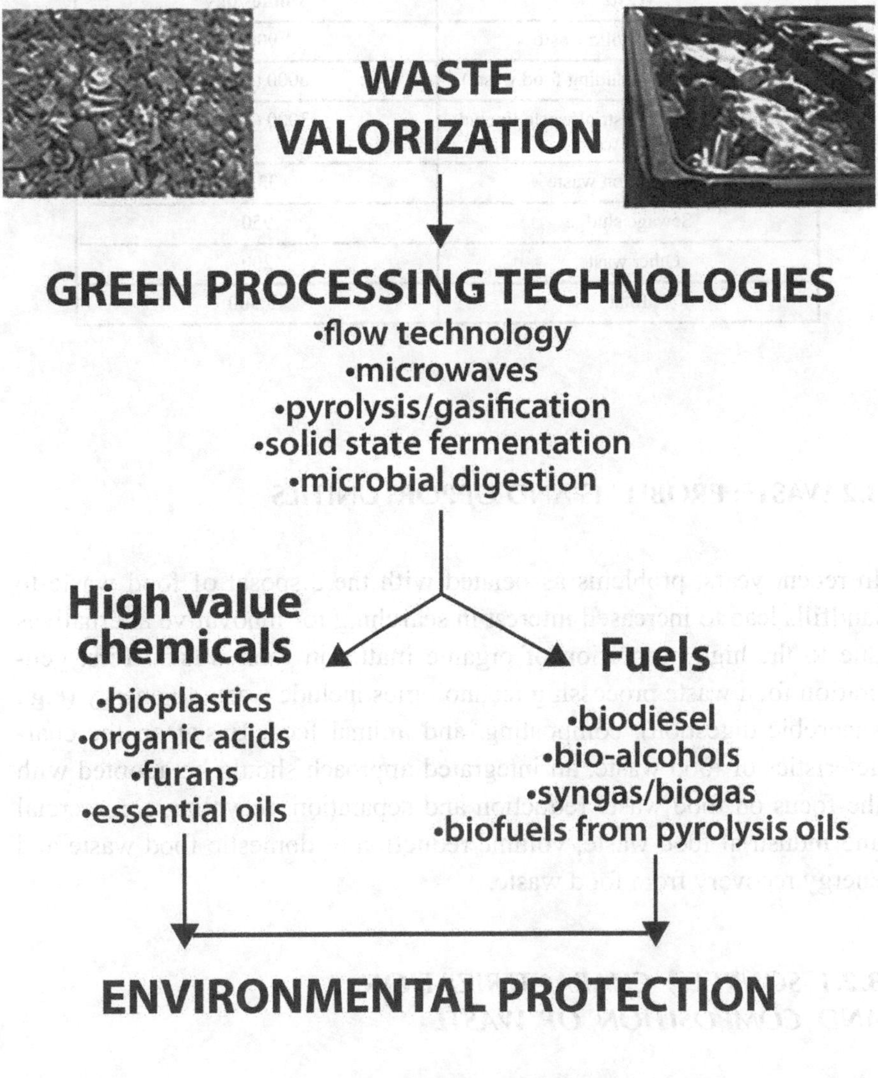

Figure 1. Valorization is essentially a concept of recycling waste into more usable industrial chemicals. Using established Green Processing technologies, various types of waste can be converted into high-value chemicals and fuels with the purpose of minimizing waste disposal volumes and eventually protecting the environment.

Table 1. Composition of waste in Hong Kong [20].

Waste	Tonnes/day
Municipal solid waste	9000
Domestic waste (including food waste)	6000 (2550)
Commercial and industrial waste (including food waste)	3000 (1050)
Construction waste	3350
Sewage sludge	950
Other waste	200
Total	13,500

3.2 WASTE: PROBLEMS AND OPPORTUNITIES

In recent years, problems associated with the disposal of food waste to landfills lead to increased interest in searching for innovative alternatives due to the high proportion of organic matter in food waste,. First generation food waste processing technologies include waste to energy (e.g., anaerobic digestion), composting, and animal feed. Based on the characteristics of food waste, an integrated approach should be adopted with the focus on food waste reduction and separation, recycling commercial and industrial food waste, volume reduction of domestic food waste and energy recovery from food waste.

3.2.1 SOURCES, CHARACTERIZATION, AND COMPOSITION OF WASTE

The large amounts of waste generated globally present an attractive sustainable source for industrially important chemicals. Food waste including garbage, swill, and kitchen refuse [21], can be generally described as any by-product or waste product from the production, processing, distribution, and consumption of food [22].

The definition of food waste is, however, different in different countries or cities. In the European Union, food waste is defined as "any food substance, raw or cooked, which is discarded, or intended or required to be discarded." The United States Environmental Protection Agency (EPA), on the other hand, defines food waste as "Uneaten food and food preparation waste from residences and commercial establishments such as grocery stores, restaurants, and produce stands, institutional cafeterias and kitchens, and industrial sources including employee lunchrooms." In the United Nations, "Food waste" and "Food loss" are distinguished. Food losses refer to the decrease in food quantity or quality, which makes it unsuitable for human consumption [23] while food waste refer to food losses at the end of the food chain due to retailers' and consumers' behavior [24]. All in all, food waste includes not just wasted foodstuffs, but also uncooked raw materials or edible materials from groceries and wet market.

Food waste is generally characterized by a high diversity and variability, a high proportion of organic matter, and high moisture content. Table 1 summarizes some reported characteristics of food waste, indicating moisture content of 74–90%, volatile solids to total solids ratio (VS/TS) of 80–97%, and carbon to nitrogen ratio (C/N) of 14.7–36.4 [25]. Due to these properties, food waste disposal constitutes a significant problem due to the growth of pathogens and rapid autoxidation [26]. As there are already many different microorganisms in food waste, the high rate of microbial activity and the amount of nutrients in food wastes facilitate the growth of pathogens, which cause the concern for foul odor, sanitation problems, and could even lead to infectious diseases. The high moisture contents [23] also increase the cost of food waste transportation. Food waste with high lipid content is also susceptible to rapid oxidation. The release of foul-smelling fatty acids also adds difficulties to the storage of treatment of food waste (Table 2).

According to a study commissioned by the United Nations Food and Agriculture Organization (UNFAO) in 2011, 1.3 billion tonnes of food waste is generated per year and roughly one third of food produced for human consumption is lost or wasted globally. The report also noted that food waste of industrialized countries and developing countries have different characteristics. Firstly, increasingly important quantities of food waste are generated in industrialized countries as compared to volumes observed in

Table 2. Characteristics of reported domestic food waste [18].

Source	Characteristics			
	Moisture content (%)	Volatile solid/total solid (%)	Carbon/nitrogen	Country
A dining hall	80	95	14.7	Korea
University's cafeteria	80	94	NAa	Korea
A dining hall	93	94	18.3	Korea
A dining hall	84	96	NA	Korea
Mixed municipal sources	90	80	NA	Germany
Mixed municipal sources	74	90–97	NA	Australia
Emanating from fruit and vegetable, markets, household and juices centers	85	89	36.4	India

[a] NA, not available.

developing countries on a per capita basis. Figure 2 shows the per capita food loss in Europe and North America is 280–300 kg/year. In contrast, the food loss per capita in sub-Saharan Africa and South/Southeast Asia accounts for 120–170 kg/year. Also, food waste is mainly generated at retail and consumer levels in industrialized countries. Comparatively, food waste is generated in developing countries mainly at postharvest and processing levels, supported by the per capita that is, food waste generated by consumer levels in Asia is only 6–11 kg/year [27].

Interestingly, the amount of food waste generated for example in Hong Kong is staggering. Figure 3 shows an increasing trend of the food waste generated daily from 3155 tonnes in 2002 to 3484 tonnes in 2011. Although the disposal of food waste in landfills was found to be the most economical option [28], it causes numerous problems in landfill sites. As landfilling disposal generally buries and compacts waste under the ground, the decomposition of food waste produces methane, a GHG that is twenty-one times powerful than carbon dioxide (CO_2) under anaerobic

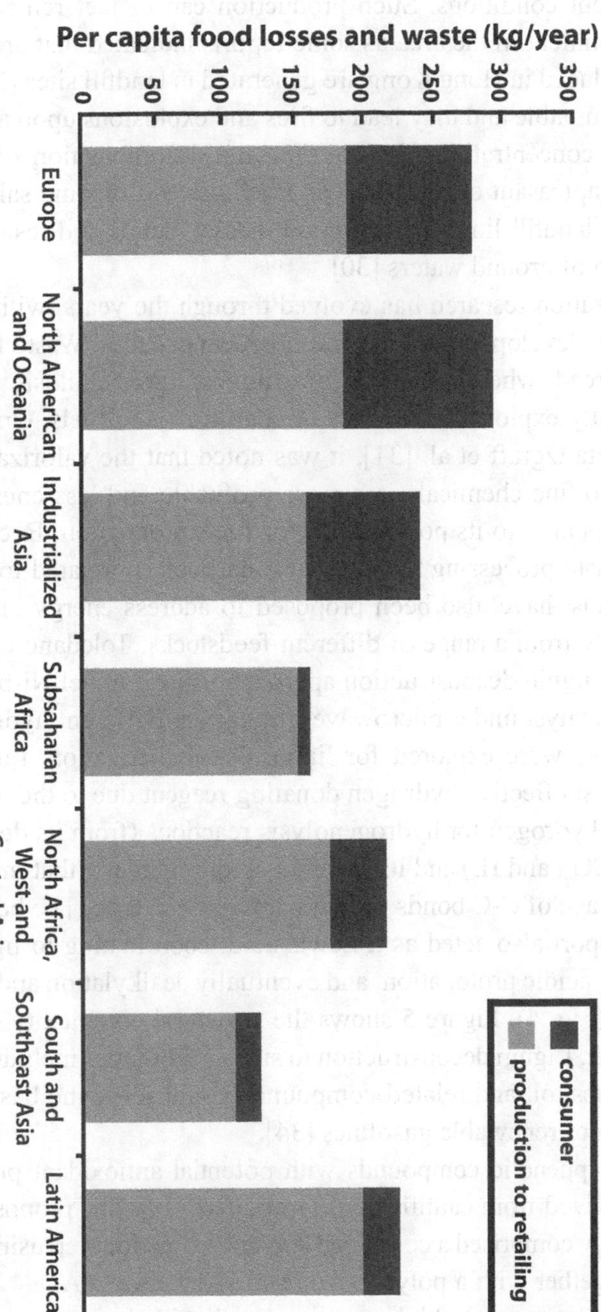

Figure 2. Per capita food losses and waste, at consumption and pre-consumption stages, in different regions [19].

environment conditions. Such production can in fact remarkably affect the environment in the area as some reports indicated that around 30% of GHG produced in Hong Kong are generated in landfill sites [29]. Methane is also flammable and may lead to fires and explosions upon accumulation at certain concentrations. In addition, the decomposition of food waste develop unpleasant odor as well as leachates and organic salts that could damaging landfill liners, leaching out heavy metals and resulting in contamination of ground waters [30].

Valorization research has evolved through the years, with many techniques and developments achieved in recent decades. Waste feedstock including bread, wheat, orange peel residues, lignocellulosic sources, etc. are currently explored as sources of chemicals and fuels. On a recent review by Pfaltzgraff et al. [31], it was noted that the valorization of food wastes into fine chemicals is a more profitable and less energy consuming as compared to its possibilities for fuels production. Because of this, related waste processing technologies, particularly related to the production of fuels, have also been proposed to address energy efficiency and profitability from a range of different feedstocks. Toledano et al. [32, 33] reported a lignin deconstruction approach using a novel Ni-based heterogeneous catalyst under microwave irradiation. Different hydrogen donating solvents were explored for lignin depolymerization, finding formic acid as most effective hydrogen donating reagent due to the efficient generation of hydrogen for hydrogenolysis reactions (from its decomposition into CO, CO_2, and H_2) and its inherent acidic character that induces acidolytic cleavage of C-C bonds in lignin at the same time. The heterogeneous acidic support also acted as a Lewis acid, coordinating to lignin thereby promoting acidic protonation, and eventually dealkylation and deacylation reactions (Fig. 4). Figure 5 shows the structural complexity of the lignin biopolymer. Lignin deconstruction to simple aromatics including syringaldehyde, mesitol, and related compounds could serve the basis for a new generation of renewable gasolines [34].

Simple phenolic compounds with potential antioxidant properties can also be derived from cauliflower by-products [35]. The proposed valorization strategy comprised a combined solvent-extraction step using an organic solvent together with a polystyrene resin (Amberlite XAD – 2) to recover most phenolics prior to high performance liquid chromatography (HPLC)

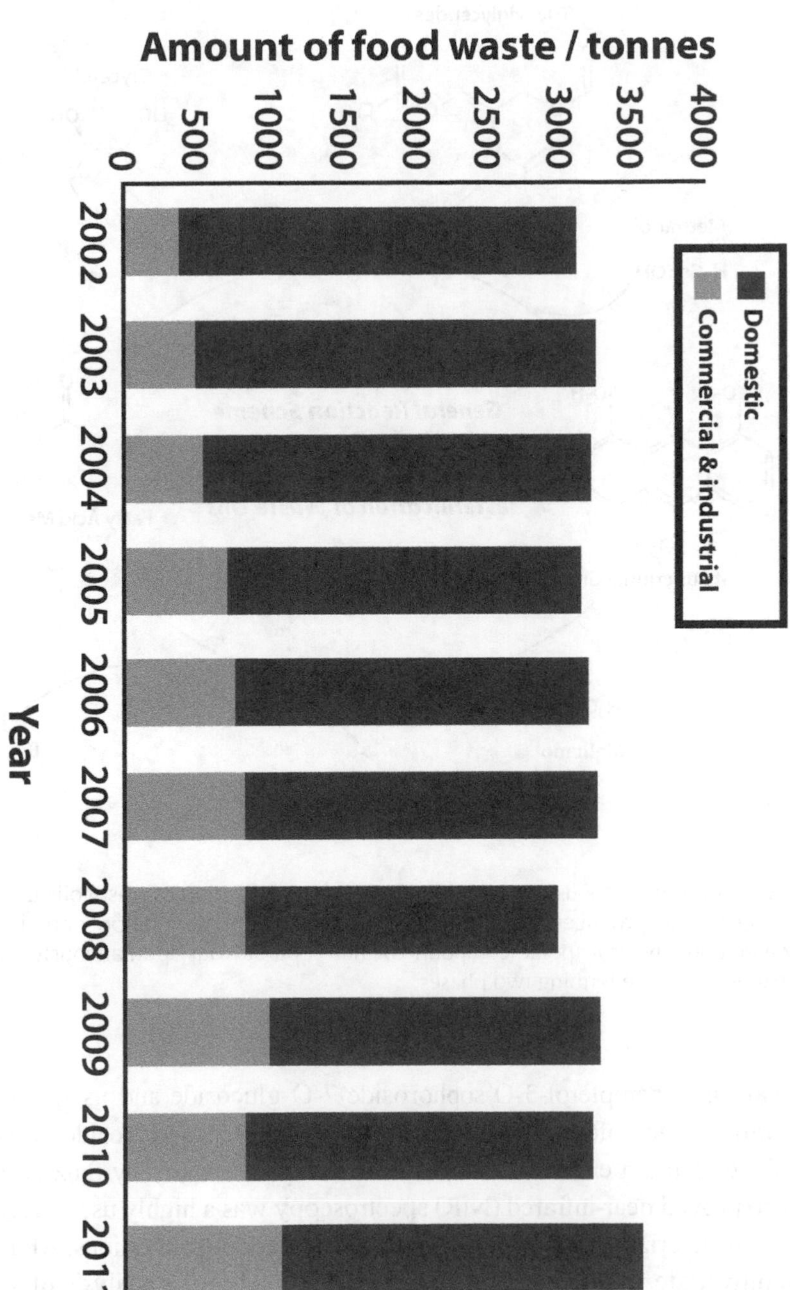

Figure 3. The amount of food waste generated daily in Hong Kong from 2002 to 2011 [25].

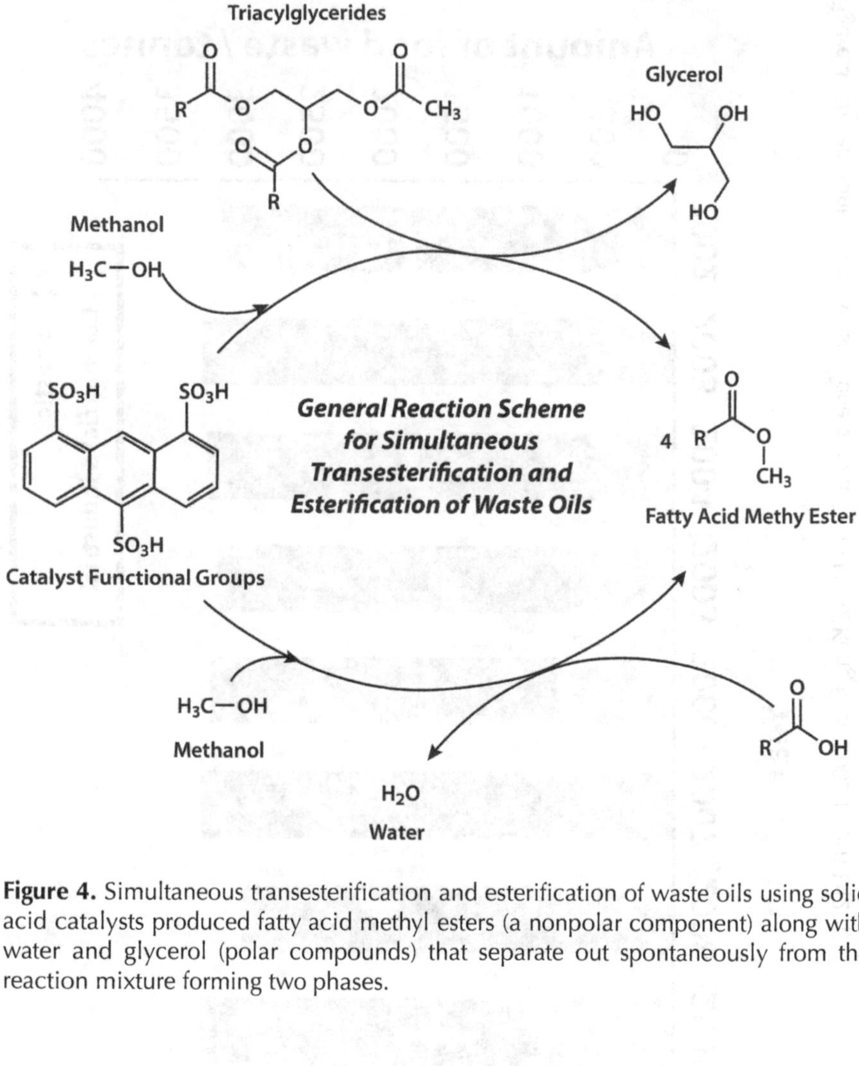

Figure 4. Simultaneous transesterification and esterification of waste oils using solid acid catalysts produced fatty acid methyl esters (a nonpolar component) along with water and glycerol (polar compounds) that separate out spontaneously from the reaction mixture forming two phases.

analysis. Kaempferol-3-O-sophoroside-7-O-glucoside and its sinapoyl derivative kaempferol-3-O-(sinapoylsophoroside)-7-O-glucoside were obtained as main extracted components. A separate study by Sáiz et al. [36] also proved near-infrared (NIR) spectroscopy was a highly useful technique to characterization online of alcohol fermentation from onions. Along with multivariate calibration, this technique can lead to the analysis of samples with complex matrices without a prior sample preparation. One approach

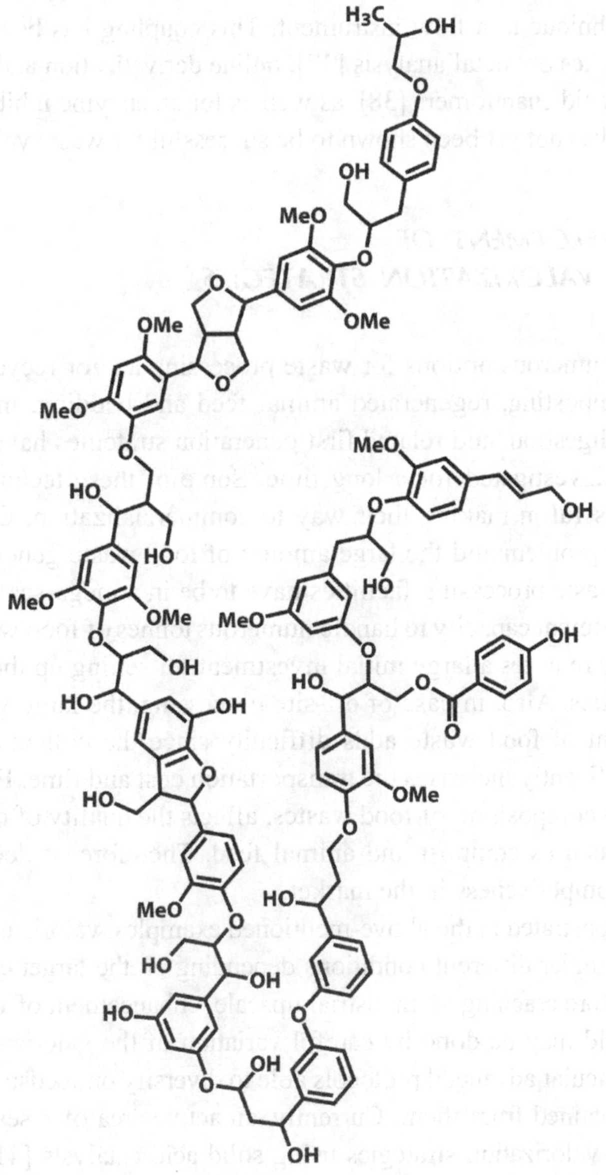

Figure 5. The structural complexicity of lignin being composed of aromatic compounds show its potential in different applications such as for fuel, and in the production of high-value chemicals (Image adapted from Stewart et al., [40]).

to a greener characterization method would be the coupling of a chromatographic technique to a flow instrument. This coupling has been shown to work in studies on metal analysis [37], online derivatization and separation of aspartic acid enantiomers [38], as well as for an enzyme inhibition assay [39], but it has not yet been shown to be successful for waste valorization.

3.2.2 DEVELOPMENT OF GREENER VALORIZATION STRATEGIES

There are numerous options for waste processing and/or recycling in the world. Composting, regenerated animal feed and bedding, incineration, anaerobic digestion, and related first generation strategies have been proposed and investigated for a long time. Some of these techniques have been successful in making their way to commercialization. Considering the storage problem and the large amount of food waste generated every day, food waste processing facilities have to be in a mega-scale size with enough treatment capacity to handle numerous tonnes of food wastes daily. It definitely requires a large initial investment for setting up the industrial scale facilities. Also, in case of off-site processing, the large volume and great weight of food waste adds difficulty since the collection of food waste significantly increases the transportation cost and time. Besides, the variation in composition of food wastes, affects the quality of regenerated products, such as compost and animal feed. Therefore, it decreases the product's competiveness in the market.

As demonstrated in the above-mentioned examples, valorization may be carried out under different conditions depending on the target components needed. Before reaching an industrial upscale, enhancement of valorization product yield may be done by careful variation of the valorization strategies, in particular advanced protocols able to diversify on feedstock and end products obtained from them. Currently, an active area of research relates to catalytic valorization strategies using solid acid catalysts [41, 42]. One example of a green protocol on valorization of waste oils to biodiesel was provided by Fu et al. [43], in which a superacid was prepared by adding a sulfuric acid solution to zirconium hydroxide powder. Under optimum reaction conditions, 9:1 MeOH/oil molar ratio, 3% (w/w) catalyst, and 4 h reaction time at 120°C, biodiesel yield reached 93.2%. Apart from metal supports functionalized with acids, carbon-based catalysts for waste valorization are also attractively developed protocols. Aside from being an easily separable reaction component, functionalized carbonaceous materials can also be recyclable. In a study by Clark et al. [44], carbonaceous materials from porous starches (Starbons®, Department of Chemistry, University of York, York, UK) functionalized with sulfonated groups were found to have a catalytic activity 2–10 times greater to those of common microporous carbonaceous catalysts in a range of chemistries including biodiesel production from waste oils and SA transformations in a fermentation broth. A separate study by Luque et al. [45] employed carbonaceous residues of biomass gasification as catalysts for biodiesel synthesis. The results showed good ester conversion yields from fatty acids to methyl esters. The above-mentioned examples demonstrate that designer catalysts can be attractive options in the valorization of a range of waste feedstocks.

A promising sustainable approach would also be the use of ionic liquid-type compounds which can be derived from renewable feedstock such as the so-called deep-eutectic solvents [46-49] and even selected designer ionic liquids. These compounds are salts in their liquid states with very unique properties such as very low vapor pressure, thermal stability, and tunability based on different applications. A study by Ruiz et al. [50] presented a $-SO_3H$ functionalized Bronsted acid ionic liquids catalyzed synthesis of an important chemical precursor such as furfural from C5 sugars under microwave heating. Furfural yield varied from 40% to 85% depending on the type of Ionic liquid used and the feedstock employed in the process. It was shown that the ionic liquid 1-(4-Sulfonylbutyl)pyridinium tetrafluoroborate produced a yield of 95% for xylose conversion and 85% for furfural. Importantly, the protocol was amenable to the utilization of a biorefinery-derived syrup enriched in C5 oligomers, from which a 40–45% of furfural yield could be derived.

A separate study by Zhang et al. [51] showed that the direct conversion of monosaccharides, and polysaccharides to 5-hydroxymethylfurfural (5-HMF) may be accomplished using ionic liquids in the presence of Germanium (IV) chloride. Yields of the reaction could go as high as 92% depending on the reaction conditions used. The mechanism proposed by the researchers indicate the role of the GeCl4 as a Lewis acid catalyst for the ring opening of the sugars, which is immediately followed by several dehydration steps to produce 5-HMF.

As alternative to these catalytic strategies, photocatalytic approaches to waste valorization could also serve the basis of innovative and highly attractive future valorization protocols. A recent review by Colmenares et al. [52] addressed the potential and opportunities of photocatalysis to convert lignin biomass into fine chemicals using designer TiO_2 nanocatalysts. These nanomaterials featuring doping agents (to lower the band gap of titania) have been shown to be effective in water splitting experiments (to form H_2 and O_2) to harness the potential of hydrogen as fuel. One of the earliest promising works of light-mediated degradation was shown by Stillings et al. [53] when they were able to degrade cellulose using Ultraviolet radiation. However, this has not been shown to be possible using visible light due to energy considerations. A photocatalytic approach to degradation may also be accomplished using functionalized graphenes (monolayers of sp^2 carbon atoms in a honeycomb lattice known to have ballistic electron transport properties). Functionalized graphenes and composites with other semiconductors have been shown to exhibit degradation properties [54, 55], but this concept has not yet been applied to waste valorization strategies.

3.2.3 RECENT LEGISLATION ON WASTE MANAGEMENT

3.2.3.1 PHILIPPINES

In Metro Manila at Philippines, almost 3.5 kg of solid waste is generated per capita every day. This amount includes food/kitchen waste, papers, polyethylene terephthalate bottles, metals, and cans. Although most Metro Manila residents do not practice the open burning of waste, a necessary waste segregation is performed for ease of collection. Being the country's capital, and one of the world's most densely populated cities, Metro Manila generates over 2400 tons of waste everyday, which equates to a government spending of Php 3.4 billion (63 Million Euros) in collection and disposal. Not much legislation is available in the Philippines in terms of waste management. Although Republic Act 9003 (Solid Waste

Management Act) has been passed last 2000, a recent 2008 study showed that it has not been properly implemented [56].

3.2.3.2 HONG KONG

One third of the food waste generated in Hong Kong come from the Commercial and Industry (C&I) sector, with the remaining percentage coming from households. In recent years, the amount of disposal of food waste from C&I sectors remarkably increased by 280% from 373 tonnes in 2002 to 1050 tonnes in 2011. It is anticipated that the food waste generated in Hong Kong will continue to rise, driven by the significant increase of the C&I food waste generation. The disposal of food waste (an organic waste which decomposes easily) to landfills is not sustainable, as it leads to rapid depletion of the limited landfill space. From the 2013 Policy Address by the Office of the Chief Executive in Hong Kong [57], there was a special emphasis on "Reduction of Food Waste" as stated in Section 142 below:

> "Food waste imposes a heavy burden on our landfills as it accounts for about 40% of total waste disposed of in landfills. In addition, odour from food waste creates nuisance to nearby residents. The Government has recently launched the "Food Wise Hong Kong Campaign" to mobilise the public as well as the industrial and commercial sectors to reduce food waste. We will build modern facilities in phases for recovery of organic waste so that it can be converted into energy, compost and other products." [57].

The Environmental Protection Department (EPD) has planned to develop Organic Waste Treatment Facilities (OWTF). Such facilities will adopt biological technologies—composting and anaerobic digestion to stabilize the organic waste and turn it into compost and biogas for recovery. The first phase of the OWTF will be constructed at Siu Ho Wan with a daily treatment capacity of 200 tonnes of source separated organic waste (Fig. 6). The second phase will be located at Sha Ling of North District with a daily treatment capacity of 300 tonnes of organic waste.

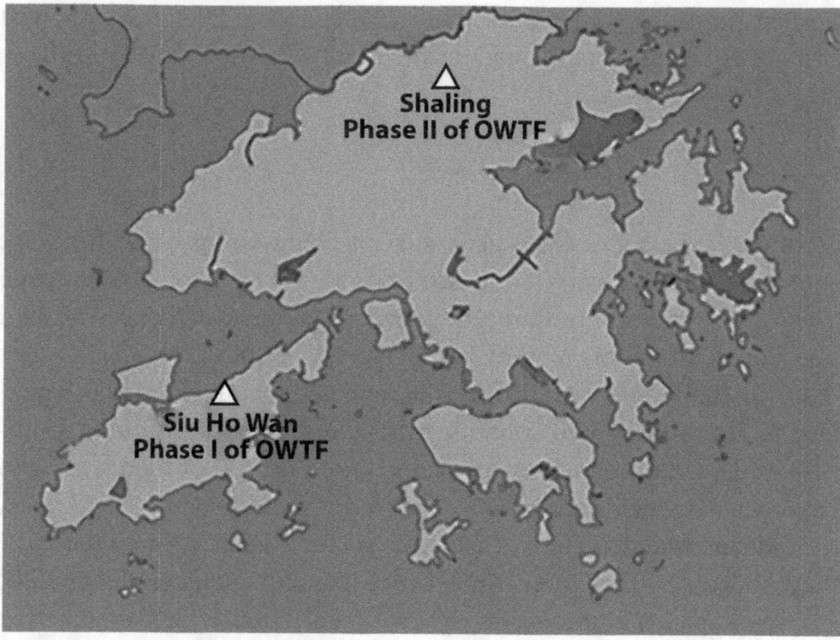

Figure 6. Map of Hong Kong indicates the location of the two organic waste treatment facility (OWTF) in Siu Ho Wan (Phase I) and Shaling (Phase II) [16].

Waste reduction at source should be the top priority so as to reduce the amount of food waste generated. Successful examples for the implementation of MSW charging scheme in Asian cities such as Taipei, Taiwan, and Seoul, South Korea could effectively reduce the total amount of MSW by 50% in 10 years [20]. These governments introduced quantity/volume-based charging scheme to create financial incentives to change public's food waste-generating behavior to achieve waste reduction at source. In addition, they introduced prepaid designated food waste bag charging system so as to achieve source separation. Food waste together with plastic bags can undergo treatment without extra separation step in the treatment facilities.

3.3 WASTE VALORIZATION STRATEGIES: CASE STUDIES

Biological treatment technologies including anaerobic digestion and composting have been reported extensively in past years. Under anaerobic digestion, biogas is generated as main product. Takata et al. [58] reported the production of 223 m^3 biogas from 1 tonne of food waste. However, Bernstad et al. [59] reported that the yield of biogas production may vary depending on the composition of waste and the existence of detergent. Numerous studies show that the lack of enough nutrients limits the ability of enzymes to digest waste [60, 21]. This can divert waste from landfill, and thus prevent the emission of GHG to the environment. Also, the solid residues can be used as compost, which can reduce the amount of used chemical fertilizers. Economically, anaerobic digestion can generate electricity on-site and may reduce energy cost. Also, it can be adopted in sewage treatment facilities, thereby eliminating transportation costs. Another way to valorize waste is by incineration for energy recovery. However, burning food waste is an energy intensive process and may remove important functional groups from the treated feedstocks. The following sections report case studies of different feedstock in different countries to illustrate the potential of waste valorization for the production of materials, chemicals, and products.

3.3.1 UTILIZATION OF BAKERY WASTE IN THE BIOTECHNOLOGICAL PRODUCTION OF VALUE-ADDED PRODUCTS

Based on the large quantities of food waste generated at Hong Kong on a daily basis, Lin et al. have been recently focused on the valorization of unconsumed bakery products to valuable products via bio-processing in collaboration with retailer "Starbucks Hong Kong". Research was initially set on the production of bio-plastics poly(3-hydroxybutyrate) (PHB) and platform chemicals (e.g., SA) via enzymatic hydrolysis of non pretreated bakery waste, followed by fungal solid state fermentation to break down carbohydrates into simple sugars for subsequent SA or PHB fermentation. In the proposed biotechnological process, bakery waste serve as the nutrient source, including starch, fructose, free amino nitrogen (FAN), and trace amount of subsidiary nutrients. The nutrient content is listed in Table 3 below.

In general, pastries have larger starch and lipid content to those of cakes; whereas cakes have higher sugar (fructose and sucrose) and protein content. Nevertheless, both types of bakery waste were proved to serve as excellent nutritional substrates for fermentative production of SA or bioplastics after hydrolysis. Our groups previously demonstrated that SA could be produced from wheat-based renewable feedstock [63-65] and bread waste [66] via fermentation. Similarly, production of biopolymers from various types of food industrial waste and agricultural crops was shown to be techno-economically feasible for replacing petroleum-derived plastics [67].

The key components in the project are illustrated in Figure 7. In the upstream processing, the bakery waste was collected from a Starbucks outlet in the Shatin New Town Plaza. A mixture of fungi comprising *Asperillus awamori* and *Asperillus oryzae* were utilized for the production of amylolytic and proteolytic enzymes, respectively. Macromolecules including starch and proteins contained in bakery waste were hydrolysed, expected to enrich the final solution in glucose and FAN. This hydrolysate was subsequently used as feedstock in a bioreaction by two different types of microorganisms (*Actinobacillus succinogenes* and *Halomonas boliviensis*) to produce (SA) and PHB, respectively.

Although food waste is a no-cost nutritional source, the application of commercial enzymes in upstream processing might not be cost-efficient. To reduce process costs, the degradation of bread and bakery waste has been previously studied [61, 66]. In these studies, *A. awamor*i and *A. oryzae* were the fungal secretors of glucoamylase protease and phosphatase as well as a range of other hydrolytic enzymes that does not require any external addition of commercial enzymes.

According to Figure 8, glucose (54.2 g/L) and FAN concentrations (758.5 mg/L) were achieved at 30% (w/v) pastry waste after enzymatic hydrolysis. On the other hand, sucrose present in cake was hydrolyzed to form 1 mole of glucose and 1 mole of fructose. The glucose (35.6 g/L), fructose (23.1 g/L), and FAN concentrations (685.5 mg/L) were achieved

Table 3. Bakery waste composition (per 100 g) [61, 62].

Content	Pastry	Cake	Wheat bran
Moisture	34.5 g	45.0 g	N/A
Starch (dry basis)	44.6 g	12.6 g	N/A
Carbohydrate	33.5 g	62.0 g	15.0 g
Lipids	35.2 g	19.0 g	6 g
Sucrose	4.5 g	22.7 g	N/A
Fructose	2.3 g	11.9 g	N/A
Free sugar			1.5 g
Fiber	N/A	N/A	50 g
Protein (TN × 5.7) (dry basis)	7.1 g	17.0 g	14.0 g
Total phosphorus (dry basis)	1.7 g	1.5 g	N/A
Ash (dry basis)	2.5 g	1.6 g	N/A

N/A, data not available.

at 30% (w/v) cake waste. Among all, waste bread hydrolysate contained the highest glucose and FAN concentrations, which were 104.8 g/L and 492.6 mg/L, respectively. These results clearly demonstrate the potential of utilizing bakery hydrolysate as generic feedstock for fermentations.

Batch fermentations on enzymatic hydrolysates were subsequently carried out to investigate the cell growth, glucose consumption as well as SA production. Cake hydrolysate consisting an initial sugar content of 23.1 g/L glucose and 18.5 g/L fructose, and pastry hydrolysate with an initial sugar content of 44.0 g/L glucose were both utilized as fermentation feedstock. At the end of fermentation, the remaining glucose was 5.2 g/L whereas fructose was 3.7 g/L. A final SA concentration of 24.8 g/L was obtained at the end point, which corresponded to a yield of 0.8 g SA/g total sugar and a productivity of 0.79 g/L.h (Fig. 9). The overall conversion of waste cake into SA was 0.28 g/g cake.

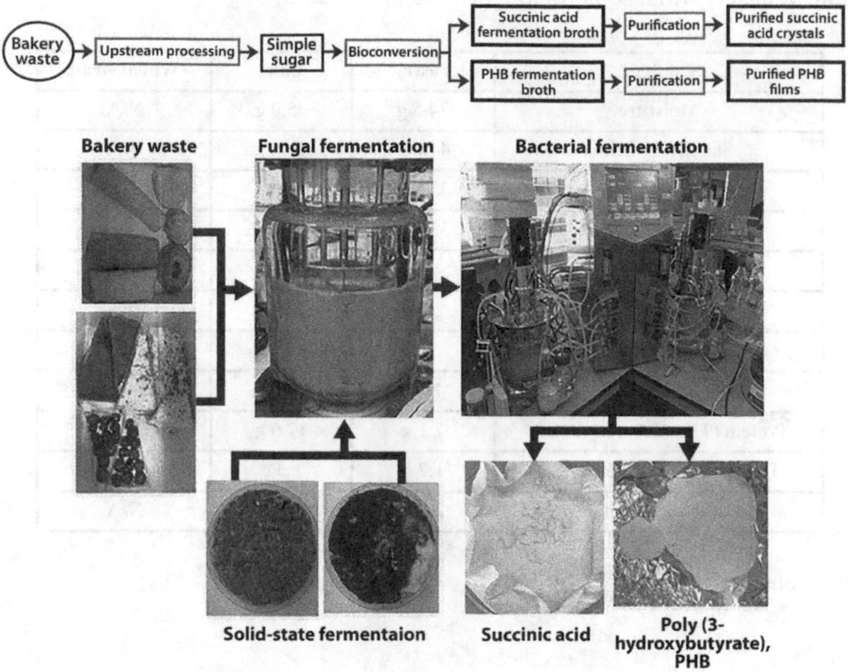

Figure 7. Flow chart of a bakery-based food waste biorefinery development, from bakery waste as raw material to succinic acid and poly(3-hydroxybutyrate), PHB as final products.

Compared with cake hydrolysates, pastry hydrolysates possessed larger concentrations of initial glucose (44.0 g/L). SA concentration continuously increased until sugar was depleted after 44 h. At the end of fermentation, the SA concentration reached 31.7 g/L, which corresponded to a yield of 0.67 g SA/g glucose and a productivity of 0.87 g/L.h.

SA production achieved from various food waste residues has been compared in Table 4. It is clear that SA yields obtained when using cake and pastry wastes as feedstock were comparable or higher to those of other food waste-derived media.

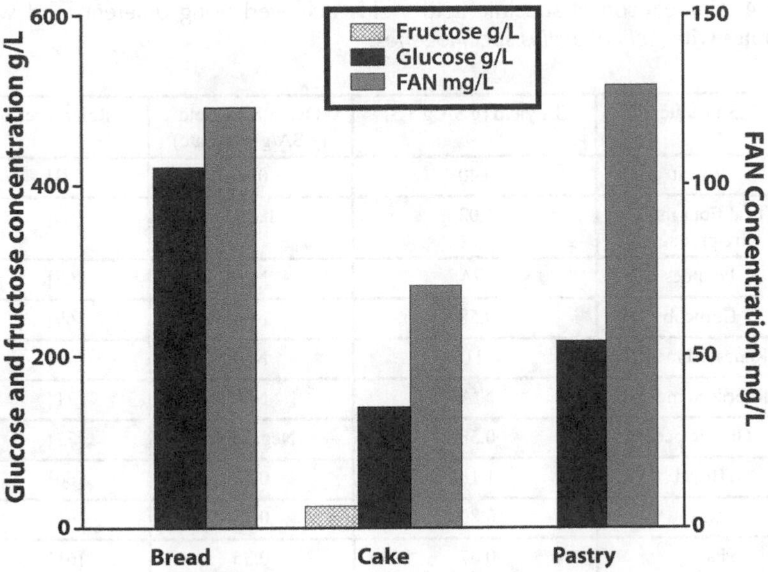

Figure 8. Sugars and FAN concentrations achieved from enzymatic hydrolysis using different bakery waste (30%, w/v) with *Aspergillus awamori* and *Aspergillus oryzae*.

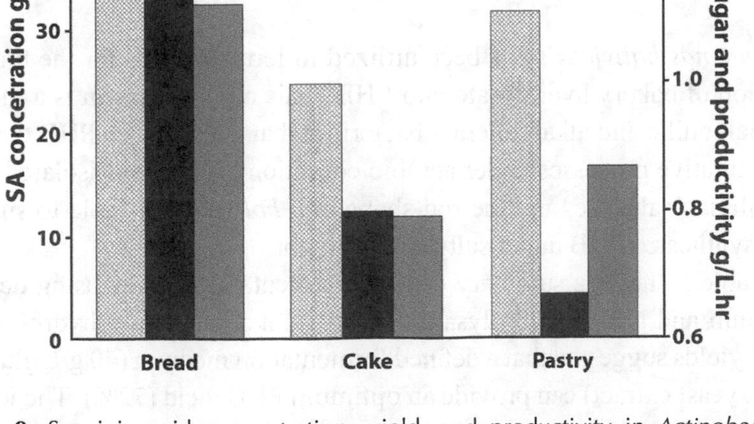

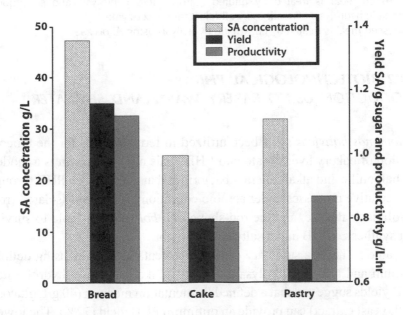

Figure 9. Succinic acid concentration, yield, and productivity in *Actinobacillus succinogenes* fermentations using different bakery hydrolysates.

Table 4. Comparison of succinic acid yields achieved using different food waste substrates with *Actinobacillus succinogenes*.

Substrate	SA yield (g SA/g TS)	Overall SA yield (g SA/g substrate)	References
Wheat	0.40	0.40	[63]
Wheat flour milling by-product	1.02	0.087	[64]
Potatoes	N/A	N/A	[68]
Corncob	0.58	N/A	[69]
Rapeseed meala	0.115	N/A	[70]
Rapeseed mealb	N/A	N/A	[71]
Orange peel	0.58	Negligible	[72]
Bread	1.16	0.55	[66]
Cake	0.80	0.28	[61]
Pastry	0.67	0.35	[61]

N/A, not available.

[a] Rapeseed meal is treated by diluted sulfuric acid hydrolysis and subsequent enzymatic hydrolysis of pectinase, celluclast, and viscozyme.

[b] Rapeseed meal is treated by enzymatic hydrolysis using *A. oryzae*.

3.3.2 BIOTECHNOLOGICAL PHB PRODUCTION USING BAKERY WASTE AND SEAWATER

Halomonas boliviensis has been utilized in fermentations for the bioconversion of bakery hydrolysate into PHB. This microorganism is a moderate halophilic and alkali tolerant bacterium that can produce PHB through fermentative processes under aerobic condition [73]. It was isolated from a Bolivian salt lake, and the rod-shaped *H. boliviensis* is able to survive and synthesize PHB under salty environment.

Table 5 shows a summary of PHB fermentation results using defined medium and bakery hydrolysates, namely cake and pastry hydrolysates. PHB yields suggested that a defined fermentation medium (40 g/L glucose, 5 g/L yeast extract) can provide an optimum PHB yield (72%). The lowest PHB yield (1–2%), as expected, was obtained under bakery hydrolysate

Table 5. The overall performance of both defined medium and bakery hydrolysate fermentation for PHB production in terms of fermentation conditions and results.

Batch no.	Fermentation medium	Fermentation mode	Feeding media	Fermentation time (h)	Glucose consumption (g)	CDW (g)	PHB production (g)	PHB content (%)
1	Defined (40 g/L glucose, 2 g/L yeast extract)	Batch	NIL	64.0	13.0	NIL	NIL	NIL
2	Defined (40 g/L glucose, 5 g/L yeast extract)	Batch	NIL	88.0	24.0	24.9	17.4	17.4
3	Defined (40 g/L glucose, 8 g/L yeast extract)	Batch	NIL	75.0	59.9	9.2	4.3	4.3
4	Pastry hydrolysate	Batch	NIL	23.5	32.8	NIL	NIL	NIL
5	Pastry hydrolysate	Fed-batch	Glucose solution	135.5	112.3	5.7	2.1	2.1
6	Pastry hydrolysate	Fed-batch	Pastry hydrolysate	67.0	208.8	38.2	3.6	3.6
7	Pastry hydrolysate	Fed-batch	Pastry hydrolysate	87.0	359.9	15.6	0.6	0.59
8	Cake hydrolysate	Fed-batch	Cake hydrolysate	63.0	200.5	11.6	2.9	2.9

CDW, cell dry weight.

fermentation media. This demonstrates that a defined medium with 40 g/L glucose and around 5 g/L yeast extract could provide sufficient nutrients for *H. boliviensis* to produce PHB efficiently.

The overall glucose consumption for defined medium fermentation in the batch mode ranged from 13 to 60 g. High initial nitrogen source could hinder PHB production by 10 times, as indicated from PHB yield obtained by defined medium. A similar effect could possibly lead to the low PHB yield observed in bakery hydrolysate fermentation. With the continuous supply of nitrogen source, *H. boliviensis* consumed glucose in a faster rate for PHB production, maintenance, and synthesis of other metabolites (six times higher overall glucose consumption with the feeding of bakery hydrolysate). *Halomonas boliviensis* synthesizes ectonie and hydroxyectonie as osmolytes as NaCl concentration increases in the cell's environment. Van-Thuoc et al. [74] reported the co-production of ectonie and PHB in a combined two-step fed-batch culture. Similarly, the formation of other primary metabolites such as ectoines in the bakery hydrolysate fermentation was observed in these studies. This consequently led to a lower PHB production when bakery hydrolysate and seawater were used as fermentation feedstocks. The highest overall yield of PHB production for the defined medium (with less glucose consumption and higher PHB production) was about 17% as compared to a rather low 3.5% observed for bakery hydrolysate.

In summary, this project is currently demonstrating the green credentials in the development of advanced food waste valorization practices to valuable products, which also include GHG reductions as well as the production of other air pollutants. Such a synergistic solution may be feasible for adoption by the Hong Kong Government as part of their strategy for tackling the food waste issue as well as for the environmentally friendly production of alternative platform chemicals and biodegradable plastics.

3.3.3 CHEMICAL VALORIZATION OF FOOD WASTE FOR BIOENERGY PRODUCTION

The valorization of waste to important chemicals can be accomplished through different approaches as discussed. Another potentially interesting approach to advanced valorization practices would be the chemical utilization of various waste raw materials for conversion into high-value products.

A case study of such integrated valorization is a recent study on the conversion of corncob residues into functional catalysts for the preparation of fatty acid methyl esters (FAME) from waste oils [75]. The design of the catalyst involved an incomplete carbonization step under air to partially degrade the lignin materials mostly present in corncobs, followed by subsequent functionalization via sulfonation to generate $–SO_3H$ acidic sites. The solid acid catalyst was then subjected to conditioning prior to its utilization in the conversion of waste cooking oil with a high content of free fatty acids (FFA) to biodiesel-like biofuels. The advantage of the designed solid acid catalyst, apart from being derived from food waste, is the possibility to conduct a simultaneous esterification of FFA present in the waste oil as well as transesterification of the remaining triglycerides also present in the oil (Fig. 4).

In this approach, the generation of two valuable products (a cheap solid acid catalyst and biodiesel-like biofuels) can be achieved starting from two food waste feedstock (corncobs and waste cooking oils). The solid acid catalysts were characterized using a range of techniques. Fourier transform infrared spectroscopy (FTIR) showed the presence of different functional groups including C=O, C-O, C-S, and aromatic C=C in the materials (Table 6). The catalytic activity of the solids also showed remarkable activity toward the conversion of waste cooking oils into biodiesel-like biofuels. A maximum of 98% yield to methyl esters could be obtained without prior purification of the oils. Importantly, kinetics of the transesterification reaction was significantly slower to those of the esterification of FFA present in the oil. Despite a low $–SO_3H$ loading (1 wt% S, 0.16 mmol/g $–SO_3H$), the catalytic activity was still high, indicating a possibly different surface functionality (Fig. 10).

The recyclability of the solid acid catalysts still, however, needs to be further optimized. Catalysts were found to deactivate quickly (after two uses) due to the aqueous promoted decomposition and hydrolysis of the sulfonated groups in the material [75]. Materials should be tested under different conditions of temperature, carbonizing atmosphere, and even pressure to improve the stability and robustness of the catalyst for the selected process. Nevertheless, this study provides a promising proof of concept of

Table 6. Summary of bands observed in IR analysis for all sulfonated samples.

Frequency of band	Corresponding functional group
1700 cm^{-1}	C=O
1597 cm^{-1}	C=C aromatic
1219 cm^{-1}	S=O
1029 cm^{-1}	C-S

the potential of an integrated valorization of various waste raw material into valuable end products and biofuels as it avoids the pretreatment of the waste oils (generally required to reduce the high FFA content to allow the conventional base-transesterification process to take place avoiding the formation of undesirable soaps and emulsions) and generates a relatively pure biofuel from a residue using an environmentally friendly and cheap solid acid catalyst.

3.3.4 TAILORED-MADE HEALING BIOPOLYMERS FROM THE MEAT INDUSTRY

The meat industry generates enormous quantities of solid waste [76]. Managing such residues entails a significant problem for the sector as many of the generated by-products and residues are prone to degradation and microbial contamination. However, an important part of these residues are rich in various added value products, which upon extraction could constitute a source of interesting revenues for these industries. Among the most promising compounds from meat industry-derived by products, we can include oily fats and collagen [77]. Valorization of the aforementioned waste fats from slaughterhouses and meat processing industries to biodiesel-like biofuels has been studied via esterification/transesterification using different types of catalysts and protocols, which entailed in some cases a pretreatment and refining of the fat [78]. In principle, these will

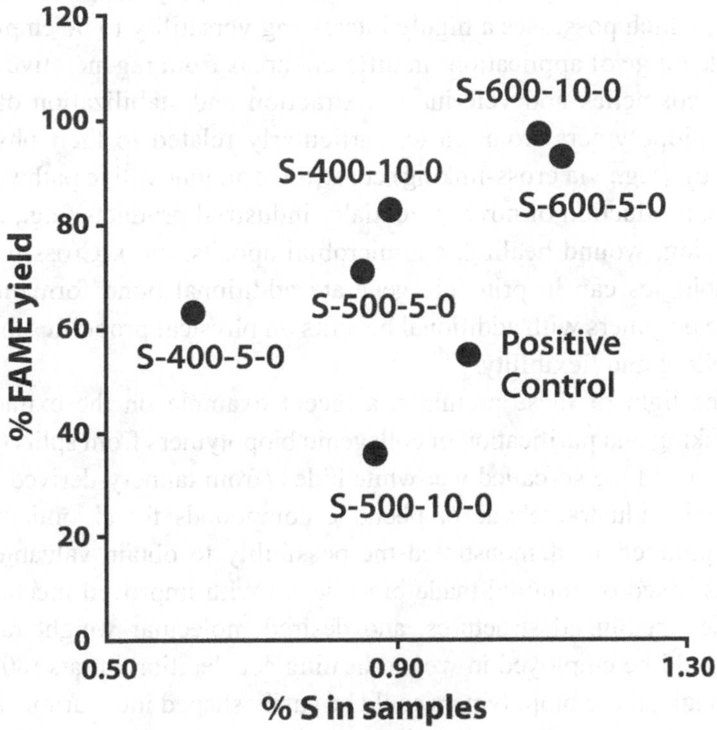

Figure 10. Plot of the FAME yield of the samples versus the %S content (A, material carbonized at 400°C for 5 h; B, carbonized at 400°C for 10 h; C, carbonized at 500°C for 5 h; D, carbonized at 500°C for 10 h; E, carbonized at 600°C for 5 h; F, carbonized at 600°C for 10 h). A higher degree of functionalization (higher %S) generally leads to improved FAME yields. Results also highlight the superior catalytic activities of sulfonated carbonaceous materials compared to blank (no conversion, data not shown) and the positive control referring to the homogeneously H_2SO_4 catalyzed reaction.

be, however, conducted in a similar way to that reported in the previously showcased study of waste cooking oils valorization.

Comparatively, collagen-containing residues (e.g., bovine hides) are increasingly important residues from the meat industry that are often derived to leather processing companies. Interestingly, the significant amounts of collagen present in such samples are not that well known [77].

Collagen is a ubiquitous and most relevant biopolymer in vertebrates [77-79], which possesses a highly interesting versatility to be employed in a wide range of applications in different areas from regenerative medicine to cosmetics and veterinary. Extraction and stabilization of collagenic biopolymers from waste, particularly related to their physical properties, (e.g., via cross-linking) constitutes an innovative pathway toward the production of novel potentially industrial products (e.g., tissue engineering, wound healing, antimicrobial aposits, etc.). Cross-linking methodologies can in principle generate additional bond formation to stabilize polymers with additional benefits on physical properties including swelling and flexibility.

In the light of these premises, a recent example on the extraction, cross-linking and purification of collagenic biopolymers from splits (pickled hides) and the so-called wet-white hides (from tannery-derived hides treated with glutaraldehyde or phenolic compounds for chromium-free leather production) demonstrated the possibility to obtain valuable end products based on tailored-made biocollagen with improved mechanical properties, stabilized structures, and desired molecular weight ranges, which could be employed in wound healing acceleration in rats [80, 81]. Interestingly, these biopolymers could be easily shaped into various forms including fibers, sponges, and/or films, paving the way to the development of potentially novel biomaterials for different biomedical applications (Fig. 11). A simple hydrolytic process was able to extract the collagen, followed by subsequent cross-linking to stable biopolymers or direct application upon purification by ultrafiltration as unguent for induced wounds in rats [80, 81].

Maximum yields of biopolymer extracted were obtained at 0.25 mm grinding size and the use of diluted acetic acid as hydrolytic agent (24 h, room temperature), for which also a minimum swelling (better biopolymer properties) was also observed. Interestingly, samples obtained from splits exhibited a better desirability to those extracted from wet-white hides. Isolated biopolymers possessed molecular weight of ca. 300 kDa, in contrast to conventional collagen derivatives for which molecular weights are usually within 15–50 kDa for hydrolysates [82] and 50–200 kDa for gelatine [83].

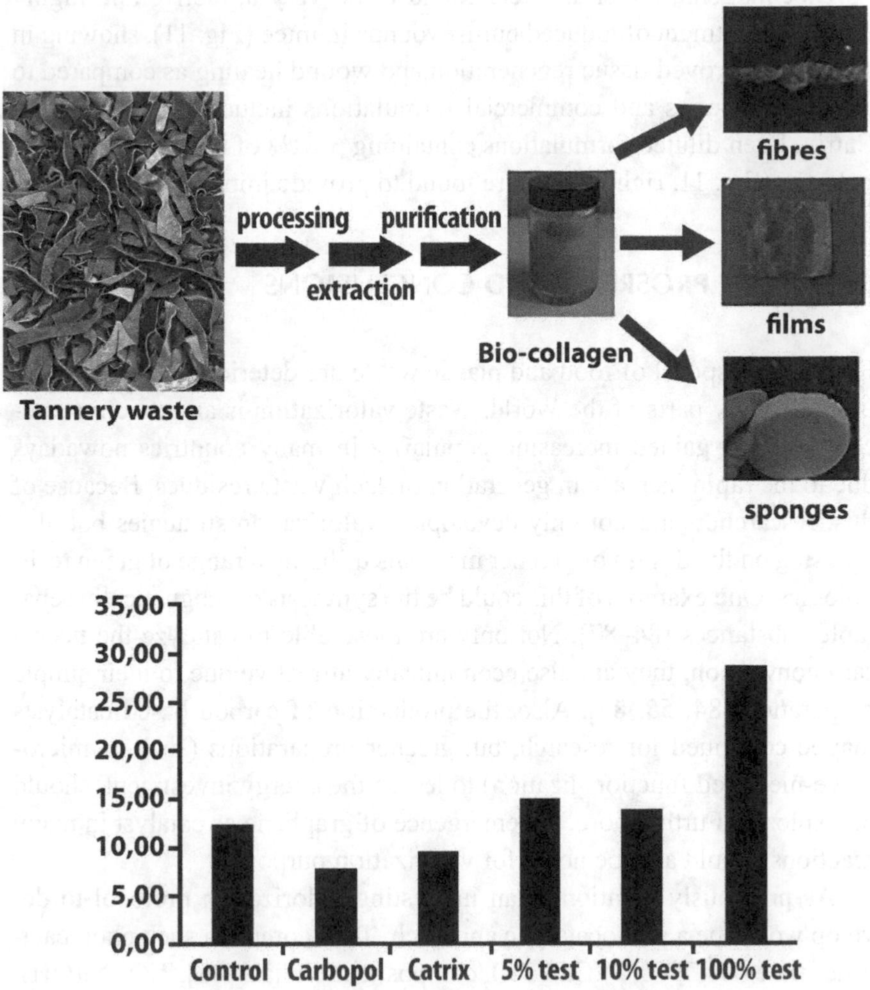

Figure 11. Meat and tannery-derived residues can be valorized to valuable collagenic biopolymers that can be formed into fibers, films, and sponges for various applications. Right plot depicts a comparison of activity in induced wounds in rats between pure and diluted collagen-extracted formulations (100%, 5% and 10% test, respectively) and a control sample (no treatment) and commercial formulations (Carbopol, Catrix).

Biocollagenic materials were found to be very attractive and highly useful in treatment of induced burns/wounds in mice (Fig. 11), showing in all cases improved tissue regeneration and wound healing as compared to untreated wounds and commercial formulations including Carbopol and Catrix. Even diluted formulations containing 5 wt% of the collagenic biopolymer (Fig. 11, right plot) were found to provide improved results.

3.4 FUTURE PROSPECTS AND CONCLUSIONS

Excessive disposal of food and plastic waste are deteriorating the landfill issue in many parts of the World. Waste valorization is an attractive concept that has gained increasing popularity in many countries nowadays due to the rapid increase in generation of such waste residues. Because of this, researchers are not only developing valorization strategies but also focusing on the design of greener materials utilizing a range of green technologies. One example of this could be the synthesis of magnetically separable substances [84–87]. Not only are these able to catalyze the necessary conversion, they are also economically attractive due to their simple preparation [84, 55, 85]. Also, the production of carbon-based catalysts maybe continued for research, but greener preparations (such as microwave-mediated functionalization) to lessen the energy investment, should be explored. Furthermore, the emergence of graphene as catalyst in many reactions should also be noted for valorization purposes.

As previously mentioned, an interesting valorization protocol to develop would be a photocatalytic approach. To accomplish such photocatalytic strategies, TiO_2, $Pt/CdS/TiO_2$ composite materials [88], $TiO_2/Ni(OH)_2$ [84] clusters may be used depending on the target, samples, and reaction conditions. Photodegradation has been shown to be possible toward many environmental pollutants such as chlorofluorocarbons [55], CO_2 [85], and NO [54], but whether these photoactive composites could degrade the stable polymeric structure of lignin/protein/carbohydrates is yet to be seen and perhaps understood. A recent study by Balu et al. [89], reports on the preparation of a TiO_2-guanidine-$(Ni,Co)Fe_2O_4$ photoactive material. The addition of the guanidine was made to lower the band gap of the material hence making it active under visible light. Testing the material to a model chemical reaction, using malic acid and the synthesized photomaterial produced simpler chemicals such as formic acid, acetic acid, and oxalic acid with a selectivity of around 80%. This study provides proof of concept that band gap engineering of semiconductors can lead to the development of photoactive materials that may be used selectively for waste valorization. A photocatalytic approach will most importantly address one of the major drawbacks of industrial valorization which is on the relatively large amounts of energy needed for processing and purification of products.

The conversion of a range of feedstock into valuable products including chemicals, biomaterials, and fuels has been demonstrated in three essentially different case studies to highlight the significant potential of advanced waste valorization strategies.

The incorporation of these and similar processes in future biorefineries for the production of value-added products and fuels will be an important contribution toward the world's highest priority target of sustainable development.

But perhaps the main and most important issue to be addressed for the sake of future generations, currently way overlooked, is society itself. The most extended perception of waste as a problem, as a residue, as something not valuable needs to give way to a general societal awareness in waste as a valuable resource. A resource, which obviously entails a significant complexity (from its inherent diversity and variability), but one that can provide at the same time an infinite number of innovative solutions and alternatives to end products through advanced valorization strategies. These will need joint efforts from a range of disciplines from engineering to (bio)chemistry, bio(techno)logy, environmental sciences, legislation, and economics to come up with innovative alternatives that we hope to see leading the way toward a more sustainable bio-based society and economy.

REFERENCES

1. Serrano-Ruiz, J. C., R. Luque, J. M. Campelo, and A. A. Romero. 2012. Continuous-flow processes in heterogeneously catalyzed transformations of biomass derivatives into fuels and chemicals. Challenges 3:114–132.

2. Glasnov, T. N., and C. O. Kappe. 2011. The microwave-to-flow paradigm: translating high-temperature batch microwave chemistry to scalable continuous-flow processes. Chem. Eur. J. 17:11956–11968.
3. PBL Netherland Environmental Assessment Agency. Trends in global CO2 emissions. http://edgar.jrc.ec.europa.eu/CO2REPORT2012.pdf (accessed 15 March 2013).
4. Mohan, D., C. U. Pittman, and P. H. Steele. 2006. Pyrolysis of wood/biomass for bio-oil: a critical review. Energy Fuels 20:848–889.
5. Heo, H. S., H. J. Park, Y.-K. Park, C. Ryu, D. J. Suh, Y.-W. Suh, et al. 2010. Bio-oil production from fast pyrolysis of waste furniture sawdust in a fluidized bed. Bioresour. Technol. 101:S91–S96.
6. Cho, M.-H., S.-H. Jung, and J.-S. Kim. 2009. Pyrolysis of mixed plastic wastes for the recovery of benzene, toluene, and xylene (BTX) aromatics in a fluidized bed and chlorine removal by applying various additives. Energy Fuels 24:1389–1395.
7. Kantarelis, E., and A. Zabaniotou. 2009. Valorization of cotton stalks by fast pyrolysis and fixed bed air gasification for syngas production as precursor of second generation biofuels and sustainable agriculture. Bioresour. Technol. 100:942–947.
8. Luque, R., J. A. Menendez, A. Arenillas, and J. Cot. 2012. Microwave-assisted pyrolysis of biomass feedstocks: the way forward? Energy Environ. Sci. 5:5481–5488.
9. Wulff, N., H. Carrer, and S. Pascholati. 2006. Expression and purification of cellulase Xf818 from Xylella fastidiosa in Escherichia coli. Curr. Microbiol. 53:198–203.
10. Atsumi, S., T. Hanai, and J. C. Liao. 2008. Non-fermentative pathways for synthesis of branched-chain higher alcohols as biofuels. Nature 451:86–89.
11. Gustavsson, L., P. Börjesson, B. Johansson, and P. Svenningsson. 1995. Reducing CO2 emissions by substituting biomass for fossil fuels. Energy 20:1097–1113.
12. Lee, S. W., T. Herage, and B. Young. 2004. Emission reduction potential from the combustion of soy methyl ester fuel blended with petroleum distillate fuel. Fuel 83:1607–1613.
13. Gielen, D. J, A. J. M. Bos, M. A. R. C. de Feber, and T. Gerlagh. Biomass for greenhouse gas emission reduction. http://www.ecn.nl/docs/library/report/2000/c00001.pdf (accessed 15 March 2013).
14. Gustavsson, L., J. Holmberg, V. Dornburg, R. Sathre, T. Eggers, K. Mahapatra, et al. 2007. Using biomass for climate change mitigation and oil use reduction. Energy Policy 35:5671–5691.
15. Chen, K., H. Zhang, Y. Miao, M. Jiang, and J. Chen. 2010. Succinic acid production from enzymatic hydrolysate of sake lees using Actinobacillus succinogenes 130Z. Enzyme Microb. Technol. 47:236–240.
16. Oliveira, L. S., and S. F. Adriana. 2009. From solid biowastes to liquid biofuels. Agriculture Issues and Policies Series: 265. Available at: http://www.demec.ufmg.br/disciplinas/eng032-BL/solid_biowastes_liquid_biofuels.pdf (accessed May 2013).
17. Toledano, A., L. Serrano, A. M. Balu, R. Luque, A. Pineda, and J. Labidi. 2013. Fractionation of organosolv lignin from olive tree clippings and its valorization to simple phenolic compounds. ChemSusChem 6:529–536.

18. Du, C., J. Sabirova, W. Soetaert, and C. S. K. Lin. 2012. Polyhydroxyalkanoates production from low-cost sustainable raw materials. Curr. Chem. Biol. 6:14–25.
19. Balu, A. M., V. Budarin, P. S. Shuttleworth, L. A. Pfaltzgraff, K. Waldron, R. Luque, et al. 2012. Valorisation of orange peel residues: waste to biochemicals and nanoporous materials. ChemSusChem 5:1694–1697.
20. Au, E. 2013. Food waste management and practice in Hong Kong in Commercial and Industrial (C&I) Food Waste Recycling Seminar, 8 February 2013, Food Education Association, The Hong Kong Polytechnic University, Hong Kong.
21. Zhang, R., H. M. El-Mashad, K. Hartman, F. Wang, G. Liu, C. Choate, et al. 2006. Characterization of food waste as feedstock for anaerobic digestion. Bioresour. Technol. 98:929–935.
22. Russ, W., and R. Meyer-Pittroff. 2004. Utilizing waste products from the food production and processing industries. Crit. Rev. Food Sci. Nutr. 44:57–62.
23. Kornegay, E. T., G. W. Vander Noot, K. M. Barth, W. S. MacGrath, J. G. Welch, and E. D. Purkhiser. 1965. Nutritive value of garbage as a feed for swine. I. Chemical composition, digestibility and nitrogen utilization of various types of garbage. J. Anim. Sci. 24:319–324.
24. Westendorf, M. L. 1996. Pp. 24–32 in The use of food waste as a feedstuff in swine diets. Proceeding of Food Waste Recycling Symp. Rutgers Coop. Ext., Rutgers Univ.-Cook College, New Brunswick, NJ.
25. Grolleaud, M. 2002. Post-harvest losses: discovering the full story. Overview of the phenomenon of losses during the post-harvest system. FAO, Agro Industries and Post-Harvest Management Service, Rome, Italy.
26. Parfitt, J., M. Barthel, and S. Macnaughton. 2010. Food waste within food supply chains: quantification and potential for change to 2050. Philos. Trans. R. Soc. Lond. B Biol. Sci. 365:3065–3081.
27. Gustavsson, J., C. Cederberg, U. Sonesson, R. van Otterdijk, and A. Meybeck. 2011. Global food losses and food waste: extent, causes and prevention. FAO, Rome, Italy.
28. Tatsi, A., and A. Zouboulis. 2002. A field investigation of the quantity and quality of leachate from a municipal solid waste landfill in a Mediterranean climate (Thessaloniki, Greece). Adv. Environ. Res. 6:207–219.
29. EPD (Environmental Protection Department of HKSAR). Monitoring of solid waste in Hong Kong 2011. https://www.wastereduction.gov.hk/chi/materials/info/msw2011tc.pdf (accessed October 2012).
30. Abu-Rukah, Y., and O. Al-Kofahi. 2001. The assessment of the effect of landfill leachate on ground-water quality—a case study El-Akader landfill site-north Jordan. J. Arid Environ. 49:615–630.
31. Pfaltzgraff, L. A., M. De bruyn, E. C. Cooper, V. Budarin, and J. H. Clark. 2013. Food waste biomass: a resource for high-value chemicals. Green Chem. 15:307–314.
32. Toledano, A., L. Serrano, J. Labidi, A. Pineda, A. M. Balu, and R. Luque. 2013. Heterogeneously catalysed mild hydrogenolytic depolymerisation of lignin under microwave irradiation with hydrogen-donating solvents. ChemCatChem 5:977–985.

33. Toledano, A., L. Serrano, and J. Labidi. 2012. Process for olive tree pruning lignin revalorisation. Chem. Eng. J. 193–194:396–403.
34. Toledano, A., L. Serrano, A. Pineda, A. A. Romero, J. Labidi, and R. Luque. 2013. Microwave-assisted depolymerisation of organosolv lignin via mild hydrogen-free hydrogenolysis: catalyst screening. Appl. Catal. B. doi: 10.1016/j.apcatb.2012.10.015
35. Llorach, R., J. C. Espín, F. A. Tomás-Barberán, and F. Ferreres. 2003. Valorization of cauliflower (Brassica oleracea L. var. botrytis) by-products as a source of antioxidant phenolics. J. Agric. Food Chem. 51:2181–2187.
36. González-Sáiz, J. M., C. Pizarro, I. Esteban-Díez, O. Ramírez, C. J. González-Navarro, M. J. Sáiz-Abajo, et al. 2007. Monitoring of alcoholic fermentation of onion juice by NIR spectroscopy: valorization of worthless onions. J. Agric. Food Chem. 55:2930–2936.
37. Dong, L.-M., X.-P. Yan, Y. Li, Y. Jiang, S.-W. Wang, and D.-Q. Jiang. 2004. On-line coupling of flow injection displacement sorption preconcentration to high-performance liquid chromatography for speciation analysis of mercury in seafood. J. Chromatogr. A 1036:119–125.
38. Cheng, Y., L. Fan, H. Chen, X. Chen, and Z. Hu. 2005. Method for on-line derivatization and separation of aspartic acid enantiomer in pharmaceuticals application by the coupling of flow injection with micellar electrokinetic chromatography. J. Chromatogr. A 1072:259–265.
39. de Boer, A. R., T. Letzel, D. A. van Elswijk, H. Lingeman, W. M. Niessen, and H. Irth. 2004. On-line coupling of high-performance liquid chromatography to a continuous-flow enzyme assay based on electrospray ionization mass spectrometry. Anal. Chem. 76:3155–3161.
40. Stewart, J. J., T. Akiyama, C. Chapple, J. Ralph, and S. D. Mansfield. 2009. The effects on lignin structure of overexpression of ferulate 5-hydroxylase in hybrid poplar. Plant Physiol. 150:621–635.
41. Sahu, R., and P. L. Dhepe. 2012. A one-pot method for the selective conversion of hemicellulose from crop waste into C5 sugars and furfural by using solid acid catalysts. ChemSusChem 5:751–761.
42. Chakraborty, R., S. Bepari, and A. Banerjee. 2010. Transesterification of soybean oil catalyzed by fly ash and egg shell derived solid catalysts. Chem. Eng. J. 165:798–805.
43. Fu, B., L. Gao, L. Niu, R. Wei, and G. Xiao. 2009. Biodiesel from waste cooking oil via heterogeneous superacid catalyst SO42−/ZrO2. Energy Fuels 23:569–572.
44. Clark, J. H., V. Budarin, T. Dugmore, R. Luque, D. J. Macquarrie, and V. Strelko. 2008. Catalytic performance of carbonaceous materials in the esterification of succinic acid. Catal. Commun. 9:1709–1714.
45. Luque, R., A. Pineda, J. C. Colmenares, J. M. Campelo, A. A. Romero, J. C. Serrano-Ruiz, et al. 2012. Carbonaceous residues from biomass gasification as catalysts for biodiesel production. J. Nat. Gas Chem. 21:246–250.
46. Abbot, A. P., R. C. Harris, K. S. Ryder, C. D'Agostino, L. F. Gladden, and M. D. Mantle. 2011. Glycerol eutectics as sustainable solvent systems. Green Chem. 13:82–90.

47. Carriazo, D., M. C. Serrano, M. C. Gutierrez, M. L. Ferrer, and F. del Monte. 2012. Deep eutectic solvents playing multiple roles in the synthesis of polymers and related materials. Chem. Soc. Rev. 41:4996–5014.
48. Zhang, Q., K. De Oliveira Vigier, S. Royer, and F. Jerome. 2012. Deep eutectic solvents: syntheses, properties and applications. Chem. Soc. Rev. 41:7108.
49. Russ, C., and B. König. 2012. Low melting mixtures in organic synthesis- an alternative to ionic liquids? Green Chem. 14:2969–2982.
50. Serrano-Ruiz, J. C., J. M. Campelo, M. Francavilla, C. Menendez, A. B. Garcia, A. A. Romero, et al. 2012. Efficient microwave-assisted production of furfural from C5 sugars in aqueous media catalysed by Brönsted acidic ionic liquids. Catal. Sci. Technol. 2:1828–1832.
51. Zhang, Z., Q. Wang, H. Xie, W. Liu, and Z. K. Zhao. 2011. Catalytic conversion of carbohydrates into 5-hydroxymethylfurfural by germanium (IV) chloride in ionic liquids. ChemSusChem 4:131–138.
52. Colmenares, J. C., R. Luque, J. M. Campelo, F. Colmenares, Z. Karpiński, and A. A. Romero. 2009. Nanostructured photocatalysts and their applications in the photocatalytic transformation of lignocellulosic biomass: an overview. Materials 2:2228–2258.
53. Stillings, R. A., and R. J. V. Nostrand. 1944. The action of ultraviolet light upon cellulose. I. Irradiation effects. II. Post-irradiation effects1. J. Am. Chem. Soc. 66:753–760.
54. Ai, Z., W. Ho, and S. Lee. 2011. Efficient visible light photocatalytic removal of NO with BiOBr-graphene nanocomposites. J. Phys. Chem. 115:25330–25337.
55. Ismail, A. A., and D. W. Bahnemann. 2011. Mesostructured Pt/TiO2 nanocomposites as highly active photocatalysts for the photooxidation of dichloroacetic acid. J. Phys. Chem. 115:5784–5791.
56. Bernardo, E. C. 2008. Solid-waste management practices of households in Manila, Philippines. Ann. NY Acad. Sci. 1140:420–424.
57. Office of the Chief Executive. 2013. Policy Address, 2013 (Office of the Chief Executive). The Hong Kong Government Special Administrative Region (HKSAR), Hong Kong. Available at http://www.policyaddress.gov.hk/2013/eng/p142.html (accessed 16 January 2013).
58. Takata, M., K. Fukushima, N. Kino-Kimata, N. Nagao, C. Niwa, and T. Toda. 2012. The effects of recycling loops in food waste management in Japan: based on the environmental and economic evaluation of food recycling. Sci. Total Environ. 432:309–317.
59. Bernstad, A., and J. la Cour Jansen. 2012. Separate collection of household food waste for anaerobic degradation – Comparison of different techniques from a systems perspective. Waste Manage. (Oxford) 32:806–815.
60. Zhang, B., L.-L. Zhang, S.-C. Zhang, H.-Z. Shi, and W.-M. Cai. 2005. The influence of pH on hydrolysis and acidogenesis of kitchen wastes in two-phase anaerobic digestion. Environ. Technol. 26:329–340.
61. Zhang, A. Y., Z. Sun, C. C. J. Leung, W. Han, K. Y. Lau, M. Li, et al. 2013. Valorisation of bakery waste for succinic acid production. Green Chem. 15:690–695.

62. Van-Thuoc, D., J. Quillaguamán, G. Mamo, and B. Mattiasson. 2008. Utilization of agricultural residues for poly(3-hydroxybutyrate) production by Halomonas boliviensis LC1. J. Appl. Microbiol. 104:420–428.
63. Du, C., S. K. C. Lin, A. Koutinas, R. Wang, P. Dorado, and C. Webb. 2008. A wheat biorefining strategy based on solid-state fermentation for fermentative production of succinic acid. Bioresour. Technol. 99:8310–8315.
64. Dorado, M. P., S. K. C. Lin, A. Koutinas, C. Du, R. Wang, and C. Webb. 2009. Cereal-based biorefinery development: utilisation of wheat milling by-products for the production of succinic acid. J. Biotechnol. 143:51–59.
65. Lin, C. S. K., R. Luque, J. H. Clark, C. Webb, and C. Du. 2012. Wheat-based biorefining strategy for fermentative production and chemical transformations of succinic acid. Biofuels Bioprod. Biorefin. 6:88–104.
66. Leung, C. C. J., A. S. Y. Cheung, A. Y.-Z. Zhang, K. F. Lam, and C. S. K. Lin. 2012. Utilisation of waste bread for fermentative succinic acid production. Biochem. Eng. J. 65:10–15.
67. García, I. L., J. A. López, M. P. Dorado, N. Kopsahelis, M. Alexandri, S. Papanikolaou, et al. 2013. Evaluation of by-products from the biodiesel industry as fermentation feedstock for poly(3-hydroxybutyrate-co-3-hydroxyvalerate) production by Cupriavidus necator. Bioresour. Technol. 130:16–22.
68. Delgado, R., A. J. Castro, and M. Vázquez. 2009. A kinetic assessment of the enzymatic hydrolysis of potato (Solanum tuberosum). LWT Food Sci. Technol. 42:797–804.
69. Yu, J., Z. Li, Q. Ye, Y. Yang, and S. Chen. 2010. Development of succinic acid production from corncob hydrolysate by Actinobacillus succinogenes. J. Ind. Microbiol. Biotechnol. 37:1033–1040.
70. Chen, K., H. Zhang, Y. Miao, P. Wei, and J. Chen. 2011. Simultaneous saccharification and fermentation of acid-pretreated rapeseed meal for succinic acid production using Actinobacillus succinogenes. Enzyme Microb. Technol. 48:339–344.
71. Wang, R., L. C. Godoy, S. M. Shaarani, M. Melikoglu, A. Koutinas, and C. Webb. 2009. Improving wheat flour hydrolysis by an enzyme mixture from solid state fungal fermentation. Enzyme Microb. Technol. 44:223–228.
72. Li, Q., J. Siles, and I. Thompson. 2010. Succinic acid production from orange peel and wheat straw by batch fermentations of Fibrobacter succinogenes S85. Appl. Microbiol. Biotechnol. 88:671–678.
73. Quillaguamán, J., R. Hatti-Kaul, B. Mattiasson, M. T. Alvarez, and O. Delgado. 2004. Halomonas boliviensis sp. nov., an alkalitolerant, moderate halophile isolated from soil around a Bolivian hypersaline lake. Int. J. Syst. Evol. Microbiol. 54:721–725.
74. Van-Thuoc, D., H. Guzmán, J. Quillaguamán, and R. Hatti-Kaul. 2010. High productivity of ectoines by Halomonas boliviensis using a combined two-step fed-batch culture and milking process. J. Biotechnol. 147:46–51.
75. Arancon, R. A., H. R. Barros Jr., A. M. Balu, C. Vargas, and R. Luque. 2011. Valorisation of corncob residues to functionalised porous carbonaceous materials for the simultaneous esterification/transesterification of waste oils. Green Chem. 13:3162–3167.

76. Cabeza, L., M. M. Taylor, G. L. DiMaio, E. Brown, W. N. Marmer, R. Carrió, et al. 1998. Processing of leather waste: pilot scale studies on chrome shavings. Isolation of potentially valuable protein products and chromium. Waste Manage. (Oxford) 18:211–218.
77. Gelse, K., E. Pöschl, and T. Aigner. 2003. Collagens—structure, function, and biosynthesis. Adv. Drug Deliv. Rev. 55:1531–1546.
78. Mata, T. M., A. A. Martins, and N. S. Caetano. 2013. Valorization of waste frying oils and animal fats for biodiesel production. Pp. 671–693 in J. W. Lee, ed. Advanced biofuels and bioproducts. Springer, The Netherlands.
79. Reis, R. L., N. M. Neves, J. F. Mano, M. E. Gomes, A. P. Marques, and H. S. Azevedo. 2008. Natural based polymers for biomedical applications. Woodhead Publishing, CRC Press, Cambridge, U.K.
80. Catalina, M., J. Cot, M. Borras, J. de Lapuente, J. González, A. M. Balu, et al. 2013. From waste to healing biopolymers: biomedical applications of bio-collagenic materials extracted from industrial leather residues in wound healing. Materials 6:1599–1607.
81. Catalina, M., J. Cot, M. Borras, J. de Lapuente, J. González, A. M. Balu, et al. 2013. From waste to healing biopolymers: biomedical applications of bio-collagenic materials extracted from industrial leather residues in wound healing. Materials 6:1599–1607.
82. Langmaier, F., P. Mokrejs, R. Karnas, M. Mládek, and K. Kolomazník. 2006. Modification of chrome-tanned leather waste hydrolysate with epichlorhydrin. J. Soc. Leather Technol. Chem. 90:29–34.
83. Brown, E., C. Thompson, and M. M. Taylor. 1994. Molecular size and conformation of protein recovered from chrome shavings. J. Am. Leather Chem. Assoc. 89:215–220.
84. Yu, J., Y. Hai, and B. Cheng. 2011. Enhanced photocatalytic H2-production activity of TiO2 by Ni(OH)2 cluster modification. J. Phys. Chem. C 115:4953–4958.
85. Liang, Y. T., B. K. Vijayan, K. A. Gray, and M. C. Hersam. 2011. Minimizing graphene defects enhances titania nanocomposite-based photocatalytic reduction of CO2 for improved solar fuel production. Nano Lett. 11:2865–2870.
86. Polshettiwar, V., R. Luque, A. Fihri, H. Zhu, M. Bouhrara, and J. M. Basset. 2011. Magnetically recoverable nanocatalysts. Cheminform 42:3036–3075.
87. Liu, J., S. Z. Qiao, Q. H. Hu, and G. Q. Lu. 2011. Magnetic nanocomposites with mesoporous structures: synthesis and applications. Small 7:425–443.
88. Daskalaki, V. M., M. Antoniadou, G. Li Puma, D. I. Kondarides, and P. Lianos. 2010. Solar light-responsive Pt/CdS/TiO2 photocatalysts for hydrogen production and simultaneous degradation of inorganic or organic sacrificial agents in wastewater. Environ. Sci. Technol. 44:7200–7205.
89. Balu, A. M., B. Baruwati, E. Serrano, J. Cot, J. Garcia-Martinez, R. S. Varma, et al. 2011. Magnetically separable nanocomposites with photocatalytic activity under visible light for the selective transformation of biomass-derived platform molecules. Green Chem. 13:2750–2758.

PART II

CASE STUDIES

CHAPTER 4

Modeling Municipal Solid Waste Management in Africa: Case Study of Matadi, the Democratic Republic of Congo

GREGORY YOM DIN AND EMIL COHEN

4.1 INTRODUCTION

A large amount of MSW in Africa still piles up on the spot, in trenches, ditches, riverbanks and roadsides, the waste burns in open air or drift by the river flow or the heavy rains. The consequences are air pollution, contamination of soil, groundwater, and rivers. In many Sub-Saharan African cities, MSW generation per capita/day ranges between 0.3 and 0.8 kilograms [1,2]. This amount is by far below the values in developed countries. Reports from OECD countries show average waste generation of 1.39 kg/capita/day [3]. The statistically significant relationship between waste generation per capita, on one hand, and gross national income and the human development index (this measures the country's achievements in a long and healthy life, knowledge and a decent standard of living), on the other hand, explains this gap [4].

All the above is valid for Matadi, the main seaport of the Democratic Republic of Congo (the DRC). In 2010- 2011, this country had the world lowest GDP per capitain the range of 200 to 230 US$ [5]. Matadi is the capital of the province Congo Central situated in the bank of the Congo River. Approximately 350,000 citizens live in Matadi within the populated area of 224 km2 (UNAID centre in Matadi). Until 2011, MSW collection system was not common practice in Matadi. The waste was buried onsite, alongside the roads, in the streets, between houses, on low ground, in man-made channels. Burning waste was implemented everywhere. Disposal or transfer stations for waste do not exist in the province. The unsolved problem of municipal waste along with other human activities causes environmental damage to the Congo River, the second largest river in the world after the Amazon River in terms of the size of its drainage and water discharge [6].

In 2008, the Government of the province appealed for a study of possible MSW management solution in Matadi referring to the following questions:

Phase 1: Planning collection and transportation of MSW; what are the needed organizational, technical and financial resources? Is it feasible and sustainable?

Phase 2: MSW treatment: is it a profitable enterprise?

The purpose of this article is to present the key elements for best performance and profitability of MSW management in a low-income city. This requires processing and analyzing the database, implementing quantitative methods and models to draw conclusions and results. We had to compose applicable methods that would express our approach to the problems and concerns of MSW management in low-income cities in Africa.

In this study we provide methods relating to the case of Matadi. We answer the above questions and conclude with the importance of the study for other low-income cities that are facing similar problems and concerns regarding MSW management.

There are a number of features that define MSW management in Africa.

a) Generally, the level of waste management is very low. Furthermore, it differs between the city zones [7-9];

b) Techno-organizational reasons: poor accessibility within the city, lack of properly designed collection route system and time schedule, inadequate equipment;

c) Inappropriate methods of finance when the budget for municipal services is scanty. This can lead to failure to assess the revenue generating capacity of municipalities and their debt-servicing requirements [10];

d) Community issues: lack of public awareness and community involvement, lack of mandatory and environment regulations and enforcement of these regulations, no master plan designed to the region or for the future city development;

e) High population growth rate and rapid urbanization worsen chronic waste management problems [11].

Waste collection coverage in urban areas in many African countries remains low compared to developing countries in other regions. In 2009, for 6 selected African countries this coverage changed in the range from 27% to 87% while for 6 selected Latin American countries it changed from 56% to 97% [12].

The private sector can help in alleviating these problems. Private companies have an important role in the MSW management in developing countries offering a means of enhancing efficiency and lowering costs, mobilizing needed investment funds, and introducing proven and cost effective technologies along with management expertise [13]. Challenges of privatizing MSW management in one of the municipalities in Ghana and the role of tax incentives are presented [14]. A planning model for MSW collection and transportation described in our article is aimed at economic evaluating a private company in a city where the income of the company is derived from waste collection fee.

In the last years, the MSW management continues to be an environmental health burden in many African cities [15]. In the article [9] planning municipal systems of collection and transportation of MSW in low-income cities in Africa are described. In [16], the statutory, financial, and physical aspects of MSW management are discussed. Transportation distances, infrastructure quality and accessibility are identified as decisive factors on waste collection considerations in the city of Yaoundé, the capital of Cameroon.

The current MSW management in major African cities is discussed in [17]. The authors offer assessments concerning MSW collection and treatment, and specify investment and operational costs needed for expansion of waste collection service for the entire population in the surveyed cities.

The review [1] summarizes and compares GHG emissions from MSW treatment plants, in Africa particularly. The authors conclude that the CDM projects have made some progress in this field in developed countries; however, African countries lag by far behind.The study [7] provides an information base to plan for the equipment required for the collection and transport of waste in a city of Ogbomoso, Nigeria.

The study is aimed at making decisions on possibilities for waste reduction through sorting and recycling, and disposal methods. It highlights a very high proportion, more than three-quarters, of organic waste in total waste flows of the city.

To the best of our knowledge, our study provides the first empirical evidence for MSW management in Matadi.

4.2 MATERIALS AND METHODS

4.2.1 CONCEPT OF THE STUDY

A few groups of demographic, waste generation, technical and economic data for modeling MSW collection and transportation, calculating GHG emissions and carbon credits, evaluating MSW treatment are used (Figure 1: data base).

These data enable to calculate the generation of MSW, its growth during the planning period, the characteristics of equipment, labor, costs, and local conditions and abilities. At first, the plan related to the range of service of 25 km for waste collection but later the range of service was expanded to 50 km. We include Boma (a major port on Congo River, population 400 thousand) and adjacent ports to increase the collected amount of waste sufficient for the plant processing capacity and treatment efficiency. We use the IPCC first order decay model to predict GHG emissions abated due to MSW treatment plant. We survey the local waste quantities, fractions and composition. To evaluate MSW treatment technologies, we use budgetary proposals of the technologies supplied by their producers, and costs and abilities of local workshops.

For the development of the comprehensive solution of MSW treatment, we integrate all the above (Figure 1: data processing, integration).

4.2.2 EVALUATION OF WASTE QUANTITY AND COMPOSITION. BASIC ASSUMPTIONS

Using estimates received from the Matadi municipality, an annual population growth rate of 3% is assumed. The growth rate of MSW generation in Matadi is defined conservatively 1.5% per year assuming a slow growth of income and small business activity in the city in the nearest years. Until now MSW was mostly burned on the spot. The municipality did not initiate any deliberate action on waste collection or treatment, and there were no available vessels or bins for waste collection in the streets. We found access roads to the residential areas and trip roads garbled and unmaintained.

In 2008-2009 we collected data of MSW generation in Matadi. We relied on a city map provided by UN Matadi headquarters. The data about the population are from the last General Census conducted in 1994. Using these sources and assuming 35% population growth rate in Matadi since 1994, we projected a total 44 thousand households in Matadi in 2011 and assumed an average of eight capita per household. We surveyed a number of public buildings, restaurants and hotels, and family food enterprises in the city. The residential area for MSW collection can be estimated as much as 35 km^2, 200 m^2 per household plus the equal additional public space (16% of the Matadi populated area).

In the pilot phase of estimation of MSW generated, additional points of interest like major markets, open burning spots of waste (Figure 2(a)), and waste disposal alongside river banks were surveyed. We also surveyed waste disposal yards that surround Government and municipal buildings, UN offices, hotels and restaurants. We prepared a small transit site for our tests and sampling (Figure 2(b)).

To determine waste production per capita, we collected and piled the waste from sixty households for three consecutive days. A distinction was

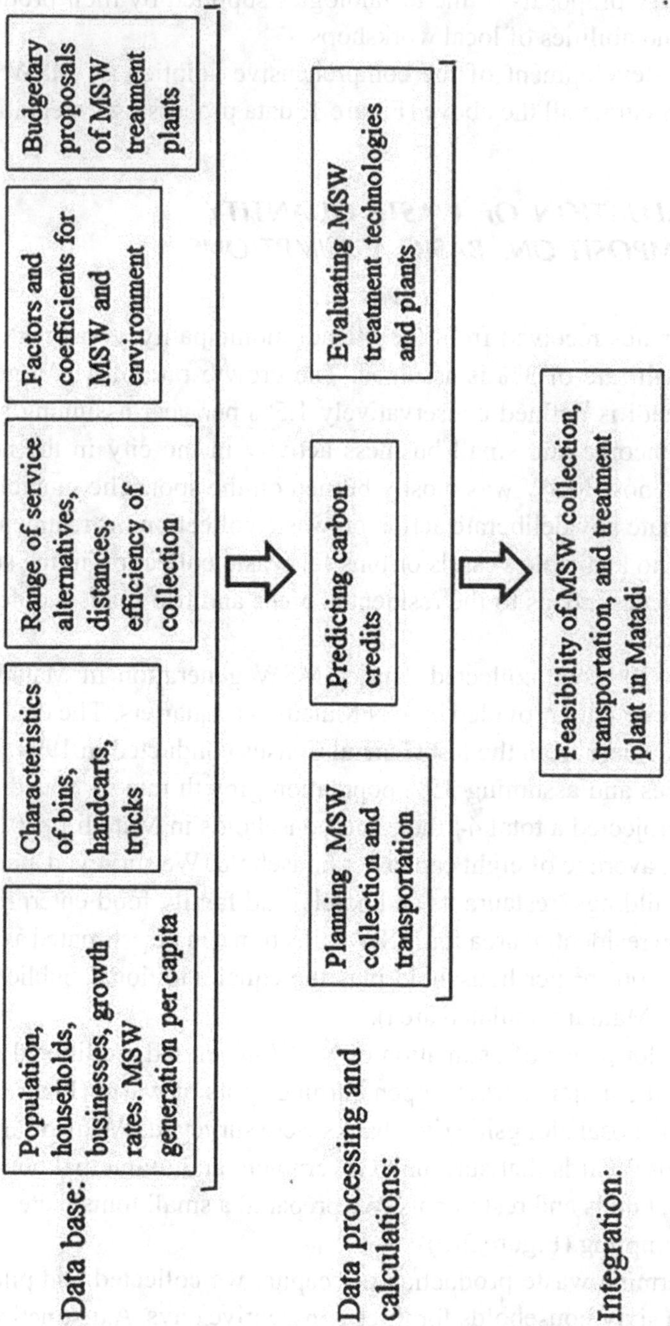

Figure 1. The concept for planning MSW management in Matadi.

made between five neighborhoods due to different life standards (workers, shop owners, small farmers). This is in line with conclusions of other researchers. The study [7] for a traditional African city in Nigeria shows that education, income and social status influence waste generation. The study [10] (Section 3.2.1) describe the experience of obtaining data concerning the waste in developing countries and note that residents of different socio-economic groups generate waste at different rates.

Five samples from each household were taken (total 5 × 60 = 300 samples). The waste from houses was weighted, divided in batches from the neighborhoods, and put on a plastic layer. The waste fractions were sampled by weight. To sort out the waste fractions, waste was sampled from all spots like riverbank, ditches, trenches, yards nearby small hotels and restaurants, and small food businesses. Based on these sources, we calculated weighted average of the waste components. We also sampled waste and residues from two major markets of the city. Seasonal variability of waste generation was assumed negligible due to the following reasons: the weather and the style of life are very stable throughout the year, and there is no tourism, colleges, seasonal business activity in the city.

The generation rate of MSW from households was estimated 91 kg per capita/year, or 0.25 kg per capita/day. The quantity of waste from small hotels, restaurants, and public buildings was estimated 3 thousand ton/year. The total generation rate of MSW was estimated 0.27 kg per capita/day, or 99 kg per capita/year. This is a low rate compared to other African low-income cities and to the estimates 0.35 - 0.65 kg per capita/day for small to medium cities (under 500,000 residents) in low-income countries ([13], Appendix A, Table 4).

We took into account that the city has no master plan available. Matadi spreads over several hills along the Congo River. The infrastructure of the city is very poor. Main roads are narrow and not maintained properly; most of them are gravel roads. Houses/barracks are aggregated in neighborhoods with low accessibility to vehicles. Therefore, planning collection and transportation of MSW involved manual labor, small bins and very few transfer stations.

From the total amount of MSW, a part of the waste, mainly from small businesses, public buildings and neighboring households is planned for collecting in big Bins B (30 m^3), and another part, mainly from

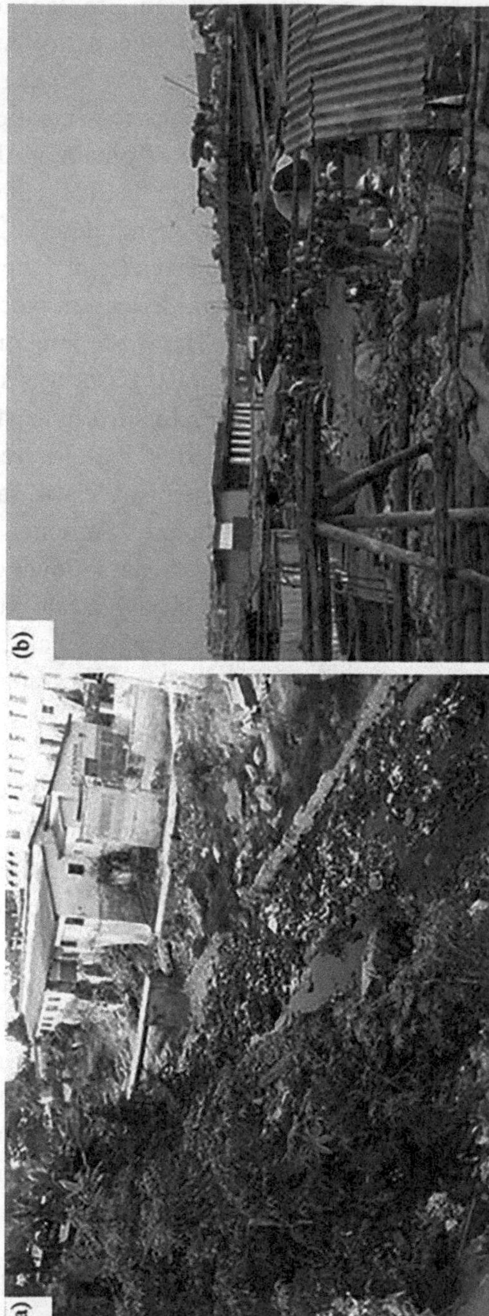

Figure 2. (a) Open burning trench in Matadi; (b) Waste sampling staff at work.

households—for collecting in Bins A (3 m^3). Handcarts with capacity of 0.25 m^3 are planned to deliver the MSW from houses to Bins A.

Fresh MSW bulk density in handcarts and bins was estimated 400 kg/m3. This is in the range of data reported by other researchers. Because it is believed that degradation of waste increases waste density we compared our estimate with data of fresh MSW. In this comparison, we took into account that MSW in developing countries has an initial density similar to that of compacted MSW from developed countries, probably due to its initial moisture content, 30% to 70% in developing countries as against 25% to 35% in developed countries [18]. In the last study, initial densities are reported 130 - 190 kg/m^3, and the density in the vehicle—400 to 550 kg/m^3 (developing countries data). According to [19], for a collection truck that compacts the waste, waste density is normally between 350 and 420 kg/m^3 (US data). In [20], the density of fresh MSW samples collected from the working face of a landfill was 515 kg/m^+ (US data).

The analysis of wet MSW in Matadi by weight showed the following composition of waste: organic matter and food residues—80%; plastic, PVC, paper, cardboards— 12.5%; other: clothes, sand and stones, glass, ceramics— 7.5%.

We compared these data with those obtained in other low-income cities. In Phnom Penh (Cambodia), the MSW composition in 1999 was close to that from Matadi: 87%, 9%, and 4%, accordingly. In 2003, after a significant increase in GDP has begun, a greater part of nonorganic components was reported, and the composition was as follows: 63%, 22%, and 15%, accordingly [21].

4.2.3 MSW COLLECTION AND TRANSPORTATION: A PLANNING MODEL AND FRAMEWORK

The spreadsheet model includes database and equations for calculating waste and machinery and equipment for its collection and transportation to the disposal site, costs, cash flow, and financial indices (Figure 3). The highlights of the model for 8 main equations are given in Appendix B.

In absence of local institutional, technical and management capacities in Matadi, it was decided to establish a company for realization, operating

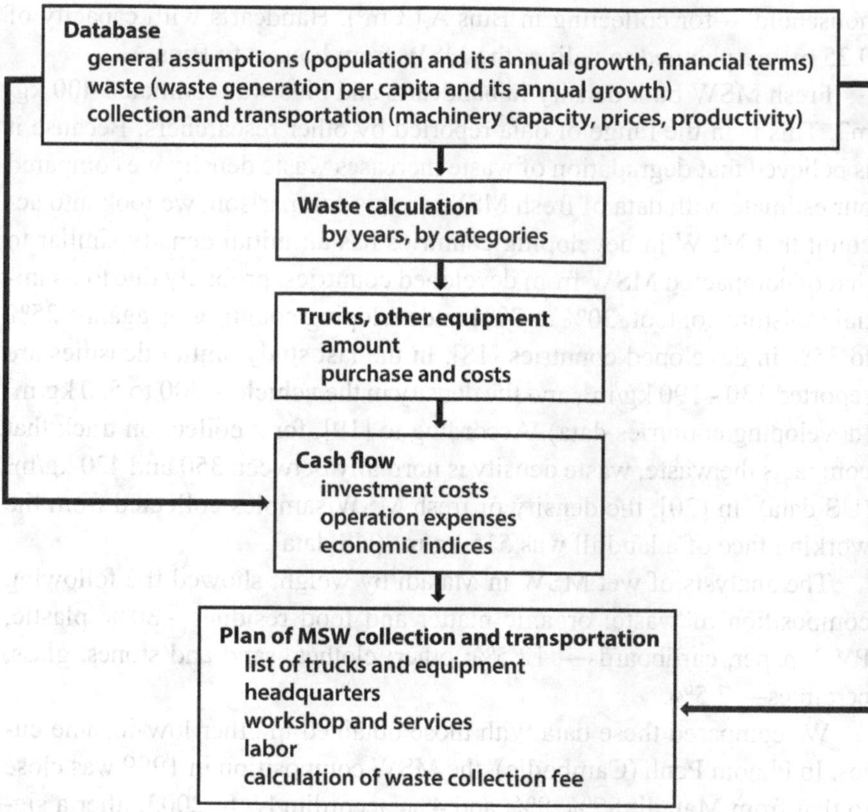

Figure 3. Flow chart of the planning model for MSW collection and transportation.

and maintaining the project. In addition, a period of chosen local team training for required tasks was planned. The output module includes all necessary information and data that justify the establishment of a company for MSW collection and transportation in Matadi ("MSW Company"). The income of the company is derived from tipping fee paid by the municipality. The model enables to determine the waste collection fee according to the planned payback period of the MSW company (five years payback period was assumed). The aims of the company are to realize the MSW collection and transportation program, to accomplish a gradual improvement in sanitation and betterment of life standard in the residential area, to employ local workers and existing abilities.

4.2.4 PREDICTION OF GHG EMISSIONS FROM DISPOSAL SITE AND CALCULATION OF CARBON CREDITS

Possible carbon credits can be predicted from GHG abated emissions in a central disposal site for MSW collected, and due to the emission savings for the generation of electricity by fossil fuel. General guidelines for this prediction, values of factors and coefficients are available in the IPCC guidelines and in the corresponding methodologies of CDM. Predicted biogas yields from non-processed MSW can be calculated based on the studied MSW composition, methane generation rates, and other constants, according to recommendations and within the normative range given by IPCC [22]. The predicted biogas yields are used for calculating carbon grants which are an important part of the economic evaluation of the MSW treatment.

The IPCC decay model enables to estimate emissions of methane from solid waste disposal sites using the first order decay model [23].

Using this model degradable organic carbon can be calculated, and on this basis, the baseline GHG emissions from the untreated MSW are predicted. The model is formulated as follows:

$$D_{t=} D_{t0} \exp(-kt) \quad (1)$$

where t is the time in years, D_t is the mass of degradable organic carbon at time t, t_0=0 at the start of the reaction, and k is the reaction constant [22].

We assume that 80% of the GHG emission would be abated due to the waste treatment plant. The abated emissions can be converted to CER's (Carbon Emission Reduction) equivalent unit (CO_2eq) used in the financial calculations of CDM.

Additional amount of CERs can be predicted due to emissions abated for the electricity generation by fossil fuel. The plant in Matadi will begin with producing 14 thousand of kWth of electricity. The needed grid emission factor can be calculated from the data published by major producers of electricity in African countries [24]. According to this publication, there was 224.7 million tons of CO_2 emissions related to sales of 218,591 GWh in 2010 by this company. It allows calculating a grid emission factor as follows: 224.7/218,591=0.001028 (tCO_2e/kWh). The value of CERs is determined in the international carbon finance bazaar. In the period studied, the price fluctuated around 13 US$/ton CO_2eq.

In [25] several different waste emissions quantification methods are compared. The methods based on IPCC 2006 guidelines were found to be more appropriate for inventorying applications.

4.2.5 EVALUATION OF MSW TREATMENT PLANT

To evaluate the MSW treatment plant the following tasks should be performed: choosing the technology, agreement upon the tipping fee payment by the municipality to the plant owner, key data assumptions, prediction of the CERs, business planning of the plant. Characteristics of technologies considered for the plant differ by investment, economic profitability, sources of waste that can be treated, process safety and environmental aspects.

After choosing the technology the following parameters are determined and submitted to the municipality: tipping fee for MSW treatment within the plant borders, expected CERs for the predicted amount of MSW, initial amount and period of the MSW accumulation before the operation of the treatment plant, the appropriate range of service, investment requirements for the chosen range of service, local abilities and infrastructure.

The following key data are used: the price of electricity is 0.1 US$/kW (2010), a number of citizens included in the service range of MSW collection is 750,000 people, area per bin A is 100 thousand m^2.

The collected data enable to prepare a business plan and forecast the economic performance of the plant. We used one of the widely accepted methods for evaluating investment projects based on the Internal Rate of Return (IRR) and derived from the discounted Cash Flow. Among other economic indices are Earnings Before Interest and Taxes (EBIT), and payback-period.

4.3 RESULTS

4.3.1 PLAN OF COLLECTION AND TRANSPORTATION OF MSW IN MATADI

The planning model (Figure 3) was used to calculate: 1) MSW generation and collection, number of bins, handcarts and needed trucks according to MSW amount; 2) cash flow and profitability of the MSW company. The logistics, storage, headquarters, control, operation and maintenance were derived from the model.

According to the model implementation, the handcart owner will collect the waste from the houses. Once the handcart is full, the worker disposes the waste content of the handcart into the bin. One handcart is operated by two workers. The company will hire approximately 240 employees, among them 220 drivers of handcarts, truck drivers, skilled and semiskilled labor.

The poor condition of access roads and the problematic topography conditions of Matadi dictate the size and number of bins and handcarts, type of trucks and the overall efficiency of collection. The compaction truck (Type A) evacuates Bins A. The platform truck (Type B) loads Bins B. Trucks of both types will evacuate the waste to the disposal site.

Calculation of MSW generation, needed trucks and equipment, capacity of bins, handcarts and trucks were done based on the key data of waste density, waste generation, on range service for bins, and on rate of waste collection (Tables 1 and 2).

Different fee values for collection and transportation of MSW were examined. The fee value of 15$ per household per year, for example, leads to 22% profitability and IRR equal to 29% along 8 years planning period. The sensitivity analysis shows how the lower fee worsens the payback and the NPV (Figure 4).

Table 1. MSW generation, needed and purchased trucks and equipment in Matadi.

Number of year	**1**	**2**	**3**	**4**	**5**	**6**	**7**	**8**
Population, thousand	350	361	371	382	394	406	418	430
For Bins A	30	31	32	33	34	36	37	39
For Bins B	5	5	6	6	6	6	7	7
Needed:								
Bins A	297	309	320	332	345	358	372	386
Bins B	5	5	6	6	6	6	7	7
Handcarts	108	112	116	121	125	130	135	140
Trucks A	3.3	3.4	3.6	3.7	3.8	4	4.1	4.3
Trucks B	0.5	0.5	0.6	0.6	0.6	0.6	0.7	0.7
Purchase								
Bins A	298	11	12	12	13	13	13	14
Bins B	6				1			
Handcarts	297	5	5	5	5	5	5	6
Trucks A	4						1	
Trucks B	1							

Table 2. Capacity of handcarts, bins, and trucks.

Item	Capacity, m. cub.	Prices, US$	Comment
Handcart	0.25	200	2 workers needed
Bin A	3	148	for neighborhoods
Bin B	30	8,214	for businesses, public buildings
Truck A	27	139,725	with compression for Bins A
Truck B	30	120,750	without compression for Bins B

4.3.2 ESTIMATED GHG EMISSIONS

The waste will be accumulated during the first 3-4 years in the disposal site. Consequently, the phase of land filling, capping and biogas extraction will start. For this period, the GHG emissions are predicted to change from 10 to 20 thousand CO_2eq ton. A small fraction of Industrial origin waste and a high organic fraction in MSW explain the relatively large amount of recovered methane per 1 ton of MSW (Figure 5). The yearly amount of CERs grants revenues is predicted to increase from 125 to 225 thousand US$.

4.3.3 PROFITABILITY OF MSW TREATMENT PLANT

The anaerobic digestion technology was chosen for the establishment of an MSW treatment plant. The economic evaluation of the plant is based on the expanded range of service (50 km) in order to provide the sufficient amount of waste for the plant efficient operation—90 thousand ton of MSW per year. For anaerobic digestion technology, it will necessary to build a sorting facility. Due to the heavy unemployment in the city, it was decided to use hand sorting. According to the capacity of the sorting

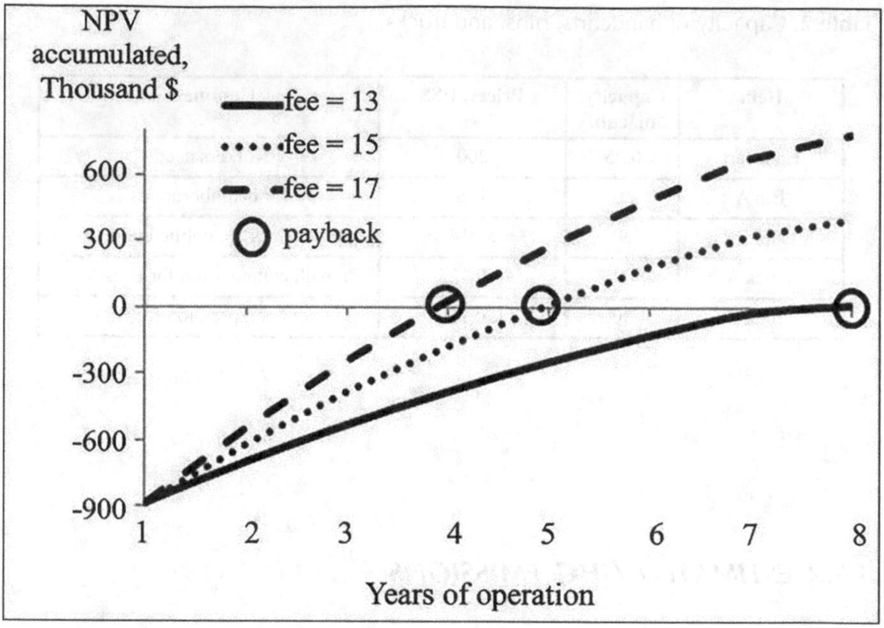

Figure 4. Profitability (in terms of NPV) of the MSW company as a function of waste collection fee.

facility (approximately 320 tons of waste daily), 32 workers are needed for its operating. Each of them will work for 280 shifts yearly sorting 10 tons of waste during a shift.

We evaluate the plant using a few major financial indices: IRR, payback period, profitability on income, and profitability on investment (ROI). The planning horizon is assumed 10 years. The first two years of the evaluation period are for planning, erection, and startup. The sources of revenues are as follows: green electricity sales to the grid (65%), tipping fee payments from the municipality—5 $/ton (19%), and CERs grants (16%). The above revenue sources are specific for a MSW treatment plant. Payback period and IRR evaluated without CERs grants are less attractive for a possible investment (Table 3). The main financial risk is the prices fall. Environmental benefits are added values for the community and life standard.

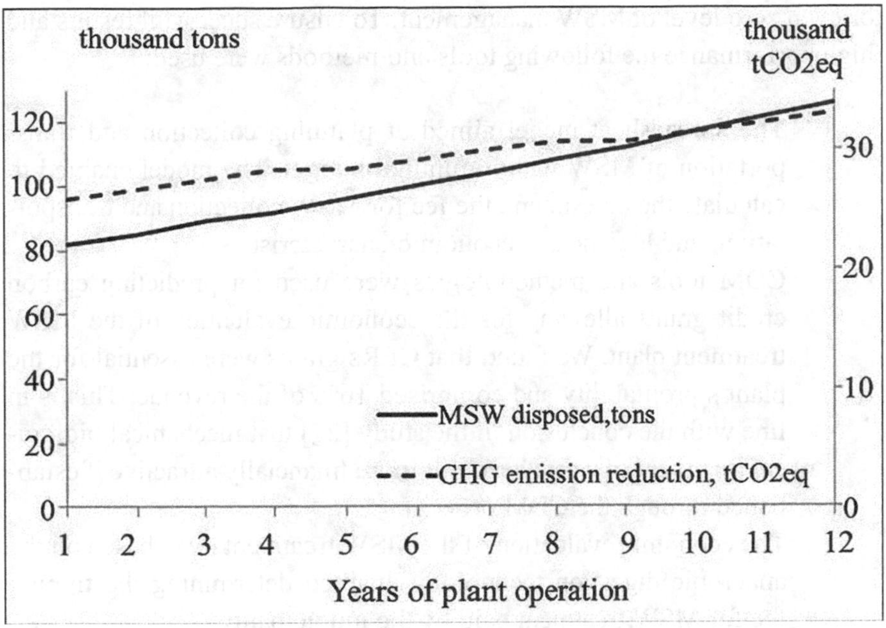

Figure 5. MSW disposal and GHG emission Reduction projected for Matadi.

4.4 CONCLUSION AND DISCUSSION

In this article, we provide the key models for comprehensive planning of collection, transportation, and MSW treatment plant in Matadi. The city of Matadi is a representative case applicable for other low-income cities in Africa where the budget for municipal services is scanty. Under these circumstances, private companies can play an important role in MSW management. Indirect evidence of this notion can be found in the article [26] where waste paper recovery is analyzed. The authors conclude that in addition to policy impacts, this sector of waste management is affected largely by economic factors such as prices and costs.

The methods used in this study can assist other low-income cities in the solution of waste problem, particularly when starting from a very low or even zero level of MSW management. To ensure successful results and high performance the following tools and methods were used:

- The spreadsheet model aimed at planning collection and transportation of MSW with a minimal budget. The model enabled to calculate the investment, the fee for MSW collection and transportation, and logistic and economic characteristics.
- CDM tools and methodologies were used for predicting carbon credit grants allowing for the economic evaluation of the MSW treatment plant. We found that CERs grants were essential for the plant's profitability and comprised 16% of the revenue. This is in line with the conclusion in the study [27] that mechanical biological treatment of waste becomes more financially attractive if established through the CDM process.
- The economic evaluation of the MSW treatment plant based on the anaerobic digestion technology enabled determining the tipping fee for MSW treatment paid by the municipality.

Topographic difficulties inflicted on collection efficiency in Matadi. Future work on improvement of MSW collection and transportation in Matadi has to include improving roads for better accessibility for vehicles, handcarts and trucks for waste collection.

The experience of the presented case study shows that the general approach for MSW management in low-income cities has to consider economic evaluation based on the joined phases of collection and transportation and MSW treatment as one project. The first phase of collection and transportation is necessary but cannot stand alone due to economic aspects and environmental regulations.

The important reason for the plant projected profitability is expansion of service range of the MSW collection from 25 to 50 km. This is possible due to the neighboring city of Boma. The second reason is a high enough collection fee as shown in the sensitivity analysis. As for demand for the electricity produced in the plant, in the present energy situation in the DRC using landfill gas can become a viable source of energy [28]. Other factors of profitability are low expenses, particularly, low salaries and taxes. On the other hand, the sold electricity price and tipping fee for

Table 3. Highlights of the MSW treatment plant evaluation.

Investment, thousand. euro	**7071**
Thousand euro per year (year 5):	
revenue	2193
profit before interest, taxes (EBIT)	1285
Profitability for EBIT (ex. year 5):	
on income	60%
on investment (ROUI)	18%
Payback period discounted, years	10
Internal rate of return (IRR, 10 years planning horizon)	17%
Without revenue from CERs:	
payback period discounted, years	13
internal rate of return	11%

MSW treatment paid by the municipality are low comparing the developed countries practice. The payback period is long (10 years including 2 years of erection and running) because of the high investment costs of the imported equipment.

REFERENCES

1. E. Friedrich and C. Trois, "Quantification of Greenhouse Gas Emissions from Waste Management Processes for Municipalities—A Comparative Review Focusing on Africa," Waste Management, Vol. 31, No. 7, 2011, pp. 1585- 1596. doi:10.1016/j.wasman.2011.02.028
2. T. Karak, R. M. Bhagat and P. Bhattacharyya, "Municipal Solid Waste Generation, Composition, and Management: The World Scenario," Critical Reviews in Environmental Science and Technology, Vol. 42, No. 15, 2012, pp. 1509- 1630. doi:10.1080/10643389.2011.569871
3. OECD, "Municipal Waste," In: OECD, Ed., OECD Factbook 2010: Economic, Environmental and Social Statistics, OECD Publishing, 2010, pp. 172-173.

4. D. C. Wilson, L. Rodic, A. Scheinberg, C. A. Velis and G. Alabaster, "Comparative Analysis of Solid Waste Management in 20 Cities," Waste Management and Research, Vol. 30, No. 3, 2012, pp. 237-254. doi:10.1177/0734242X12437569
5. World Development Indicators, "The World Bank," 2012. http://data.worldbank.org/data-catalog
6. A. Coynel, P. Seyler, H. Etcheber, M. Meybeck and D. Orange, "Spatial and Seasonal Dynamics of Total Suspended Sediment and Organic Carbon Species in the Congo River," Global Biogeochemistry Cycles, Vol. 19, No. 4, 2005. doi:10.1029/2004GB002335
7. A. Abel, "An Analysis of Solid Waste Generation in a Traditional African City: The Example of Ogbomoso, Nigeria," Environment and Urbaniztion, Vol. 19, No. 2, 2007, pp. 527-537. doi:10.1177/0956247807082834
8. R. Oyoo, R. Leemans and A. P. J. Mol, "Future Projections of Urban Waste Flows and Their Impacts in African Metropolises Cities," International Journal of Environment Research, Vol. 5, No. 3, 2011, pp. 705-724.
9. N. Regassa, R. D. Sundaraa and B. B. Seboka, "Challenges and Opportunities in Municipal Solid Waste Management: The Case of Addis Ababa City, Central Ethiopia", Journal of Human Ecology, Vol. 33, No. 3, 2011, pp. 179-190.
10. M. Coffey and A. Coad, "Collection of Municipal Solid Waste in Developing Countries," 2nd Edition, UN-Habitat, Nairobi, 2010.
11. C. Ezeah and C. Roberts, "Analysis of Barriers and Success Factors Affecting the Adoption of Sustainable Management of Municipal Solid Waste in Nigeria," Journal of Environment Management, Vol. 103, 2012, pp. 9-14. doi:10.1016/j.jenvman.2012.02.027
12. A. Scheinberg, D. C. Wilson and L. Rodic, "Solid Waste Management in the World's Cities," 3rd Edition, UNHabitat's State of Water and Sanitation in the World's Cities Series, Earthscan for UN-Habitat, London and Washington DC, 2010.
13. S. Cointreau-Levine and A. Coad, "Guidance Pack on Private Sector Participation in Municipal Solid Waste Management," SKAT, St. Gallen, 2000.
14. A. K. Yahaya and O. S. Ebenezer, "Challenges of Privatizing Waste Management in the in Wa Municipality of Ghana: A Case of Zoomlion Ghana Limited," Journal of Environment and Earth Science, Vol. 2, No. 11, 2012, pp. 68-79.
15. M. Oteng-Ababio, J. E. Melara Arguello and O. Gabbay, "Solid Waste Management in African Cities: Sorting the Facts from the Fads in Accra, Ghana," Habitat International, Vol. 39, 2013, pp. 96-104. doi:10.1016/j.habitatint.2012.10.010
16. L. Parrot, J. Sotamenou and B. K. Dia, "Municipal Solid Waste Management in Africa: Strategies and Livelihoods in Yaoundé, Cameroon," Waste Management, Vol. 29, No. 2, 2009, pp. 986-995. doi:10.1016/j.wasman.2008.05.005
17. C. Collivignarelli, M. Vaccari, V. Di Bella and D. Giardina, "Techno-Economic Evaluation for the Improvement of MSW Collection in Somaliland and Puntland," Waste Management and Research, Vol. 29, No. 5, 2011, pp. 521-531. doi:10.1177/0734242X10384431
18. N. D. Dixon and R. V. Jones, "Engineering Properties of Municipal Solid Waste," Geotextiles and Geomembranes, Vol. 23, No. 3, 2005, pp. 205-233. doi:10.1016/j.geotexmem.2004.11.002

19. P. A. Vesilind and W. A. Worrell, "Solid Waste Engineering," 2nd Edition, Cengage Learning, Stamford, 2011.
20. K. R. Reddy, J. Gangathulasi, H. Hettiarachchi and J. Bogner, "Geotechnical Properties of Municipal Solid Waste Subjected to Leachate Recirculation," In: M. V. Khire, A. N. Alshawabkeh and K. R. Reddy, Eds., GeoCongress: Geotechnics of Waste Management and Remediation (GSP 177), American Society of Civil Engineers 2008, pp. 144-151.
21. B. Seng, H. Kaneko, K. Hirayama and K. KatayamaHirayama, "Municipal Solid Waste Management in Phnom Penh, Capital City of Cambodia," Waste Management and Research, Vol. 29, No. 5, 2011, pp. 491-500. doi:10.1177/0734242X10380994
22. Intergovernmental Panel on Climate Change, "Intergovernmental Panel on Climate Change Guidelines for National Greenhouse Gas Inventories," IGES, Japan, 2006.
23. X. F. Lou and J. Nair, "The Impact of Landfilling and Composting on Greenhouse Gas Emissions—A Review," Bioresourse Technology, Vol. 100, No. 16, 2009, pp. 3792- 3798. doi:10.1016/j.biortech.2008.12.006
24. Eskom, "Integrated Report 2010: On the Path to Recovery," 2013. www.eskom.co.za
25. E. A. Mohareb, H. L. MacLean and C. A. Kennedy, "Greenhouse Gas Emissions from Waste Management—Assessment of Quantification Methods," Journal of the Air and Waste Management Association, Vol. 61, No. 5, 2011, pp. 480-493. doi:10.3155/1047-3289.61.5.480
26. C. Berglund and P. Soderholm, "An Econometric Analysis of Global Waste Paper Recovery and Utilization," Environmental and Resource Economics, Vol. 26, No. 3, 2003, pp. 429-456. doi:10.1023/B:EARE.0000003595.60196.a9
27. R. Couth and C. Trois, "Sustainable Waste Management in Africa through CDM Projects," Waste Management, Vol. 32, No. 11, 2012, pp. 2115-2125. doi:10.1016/j.wasman.2012.02.022
28. W. N. Mbav, G. Coppez, S. Chowdhury and S. P. Chowdhury, "Energy Production from Landfill Gases in African Countries," Power System Technology (POWERCON), 2010 International Conference, 24-28 October 2010, pp. 1-8.
29. J. Okot-Okumu and R. Nyenje, "Municipal Solid Waste Management under Decentralisation in Uganda," Habitat International, Vol. 35, No. 4, 2011, pp. 537-543. doi:10.1016/j.habitatint.2011.03.003
30. J. N. Fobil, N. A. Armah, J. N. Hogarh and D. Carboo, "The Influence of Institutions and organisations on Urban Waste Collection Systems: An Analysis of Waste Collection System in Accra, Ghana (1985-2000)," Journal of Environmental Management, Vol. 86, No. 1, 2008, pp. 262-271. doi:10.1016/j.jenvman.2006.12.038
31. S. Kathiravale and M. N. Muhd Yunus, "Waste to Wealth," Asia Europe Journal, Vol. 6, No. 2, 2008, pp. 359-371. doi:10.1007/s10308-008-0179-x
32. J. O. Babayemi and K. T. Dauda, "Evaluation of Solid Waste Generation, Categories and Disposal Options in Developing Countries: A Case Study of Nigeria," Journal of Applied Sciences & Environmental Management, Vol. 13, No. 3, 2009, pp. 83-88.

33. A. M. Mshandete and W. Parawira, "Biogas Technology Research in Selected Sub-Saharan African Countries—A Review," African Journal of Biotechnology, Vol. 8, No. 2, 2009, pp. 116-125.
34. T. Getahun, E. Mengistie, A. Haddis, F. Wasie, E. Alemayehu, D. Dadi, T. Van Gerven and B. Van der Bruggen, "Municipal Solid Waste Generation in Growing Urban Areas in Africa: Current Practices and Relation to Socioeconomic Factors in Jimma, Ethiopia," Environmental Monitoring and Assessment, Vol. 184, No. 10, 2011, pp. 6337-6345. doi:10.1007/s10661-011-2423-x

CHAPTER 5

Solid Waste Management in Minna, North Central Nigeria: Present Practices and Future Challenges

PETER ADEREMI ADEOYE, MOHAMMED ABUBAKAR SADEEQ, JOHN JIYA MUSA, AND SEGUN EMMANUEL ADEBAYO

5.1 INTRODUCTION

Municipal solid waste management (MSWM) will continue to be a major challenge facing countries all over the world. Especially for developing countries, where the amount of municipal solid waste (MSW) has increased greatly due to rapid increase in urban population (Adebayo et al, 2006). Meanwhile, with limited resources, only basic technologies for treatment and disposal, and deficient enforcement of relevant regulations, serious problems remain for MSWM in developing countries, especially in regard to safe disposal. The progress of modern civilization and the associated increase in population worldwide has contributed significantly to the increase in the quantity and variety of waste generated (Anikwe and Nwobodo, 2002). The increase in consumption of resources has resulted in large amounts of solid waste from domestic activities and can lead to significant threats to human health. Improper management of solid waste has serious environmental and health consequences, their environmental effects include pollution of surface and subsurface waters, unpleasant odours, pest infestations, and gas explosions (Ayo and Mohammed, 2010). Due to inadequate waste disposal, surface and groundwater are contaminated by leachate and the air is polluted by burning of waste or uncontrolled release of methane from anaerobic waste decomposition (Sha'Ato et al, 2007). The hazards associated with improper solid waste disposal and the associated environmental health impact should therefore be of utmost concern to waste management experts. If waste pollution continues unchecked, it may lead to unprecedented health consequences (Chen and Fujita, 2010).

Waste management is a global issue which needs maximum attention. In developing countries, waste management agencies lack the resources and trained staff to provide their rapidly growing populations with the necessary facilities and services for solid waste management to support good quality of life (Pokhrel and Viraraghavan, 2005). Within the framework of sustainable development, developing countries today face the challenge of balancing economic growth with environmental progress. The indiscriminate dumping of MSW is increasing and is compounded by a cycle of poverty, population explosion, decreasing standards of living, poor governance, and the low level of environmental awareness .Hence, these wastes are illegally disposed of onto any available space, known as Open-dumps (Izugbara and Umoh,2004). The collected waste is generally dumped on land in a more or less uncontrolled manner. Such uncontrolled waste disposal not only creates serious environmental problems and affects human and animal health, but also causes serious financial and socio-economic losses (Kalu et al, 2009). The potentials of residents to generate waste have increased in recent times due largely to accelerated urbanization, and population growth, which have elicited strong international concerns about the possible environmental, health and safety effects of living in the vicinity of these open-dumps. The only way to prevent this is to assess the level of waste generation, its management techniques and available disposal facilities. The objective of this paper is therefore to analyze some of the strengths and deficiencies in the current MSW management system

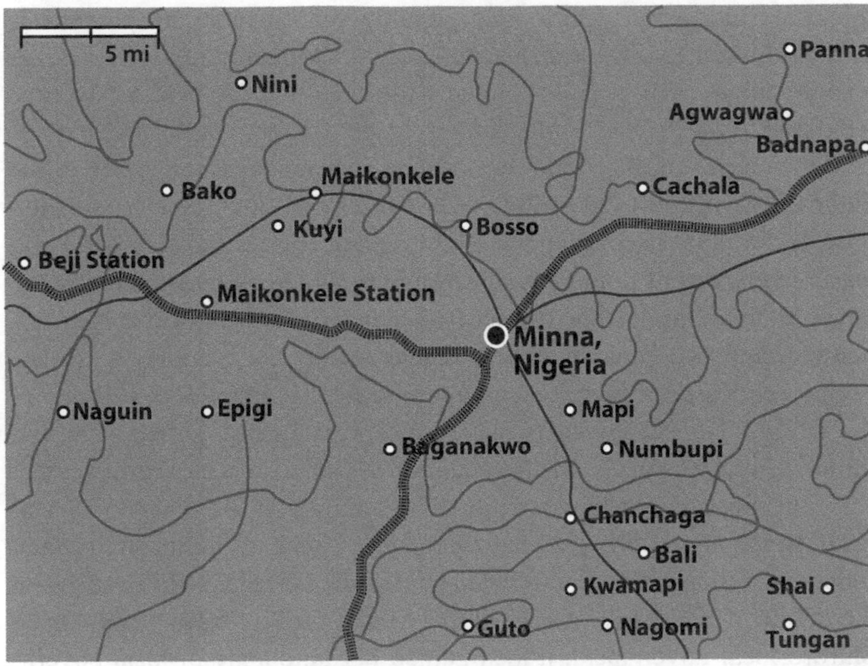

Figure 1. Map of Minna, Niger state, Nigeria.

in Minna, a fast growing city in North central Nigeria and propose feasible solutions.

5.2 MINNA

Minna, the capital city of Niger State, Nigeria has a total population of approximately 506,113. The average population density in Minna is about 3448 persons per km^2 (UNDP/NISEPA, 2009). The population growth in the city is higher than the average of the whole country because of its proximity to Abuja, the new administrative capital of the country. This shows that there may be a rapid population inflow into the city, perhaps because of job opportunities in Federal Capital Territory. However, more people means more waste, and more waste means more resources needed for waste management, the rapid population inflow should be considered in designing a waste management plan (Manaf et al, 2009). Being a small and densely populated town with a hot and dry climate, average temperature of 26.7- 35.3 0C, daily average humidity at 44.4%, and annual average rainfall of 1334 mm, Minna is potentially vulnerable to the outbreak of any infectious diseases. At the same time, rapid population growth continues to contribute to the burden of solid waste disposal. Solid waste management in the town has traditionally been undertaken by the Niger State Environmental Protection Agency (NISEPA). Minna is a multiethnic, multi-cultural, and multi-lingual society. Its economy was once exclusively based on agricultural commodities, and now it is still one of the world's largest producers of maize, sorghum, beans, rice, yam and millet. Minna was made first headquarter of Chanchaga local government since the creation of Niger State in 1976, although it still maintain its status of headquarter of Minna municipal council with all administrative and functional requirement of a full pledge local Government. However when the defunct Chanchaga local government was moved to Kuta and named Shiroro local government, it then gained her autonomy of local government in July 1989. The creation of additional local government in 1991 saw the split of chanchaga local government into the three to have Paikoro and Bosso L.G.A in addition, Figure 1. The town lies on longitude 90371N and longitude 60 33^{1}E, on geographical base of undifferentiated basement complex rock of mainly quiets and magnatile situated at the base of prominent hills in an undulating plain. The whole of Minna surrounding is very rock. The typical climate of the middle beet zone of Nigeria is a good reflection of Minna climate, with rain season starts around April and last till October. The month of September normally records highest rainfall. The mean monthly temperature is highest in March and lowest in August (UNDP/NISEPA, 2009).

5.3 WASTE GENERATION AND CHARACTERISTICS IN MINNA

Solid waste in Minna is broadly classified into three main categories: Domestic refuse (solid waste generated by households, markets, food centers and commercial premises such as hotels, restaurants, etc.). Industrial refuse (not including toxic and hazardous waste) and Institutional refuse (solid waste from various government installations like hospitals, schools and recreational facilities. Fig. 2 shows the actual amount of solid waste disposed of in tonnes in the last two decades1987-2009 in Minna city (UNDP/NISEPA, 2009).

From Fig. 2, the total solid waste in 2009 was almost three times waste generated in 1987. Domestic solid waste has increased greatly over the years, from 640tonnes to 1893tonnes in the years under review. This may be as a result of an increase in both population and per capita waste generation rate due to improved standard of living. The population was 152,603 in 1987 and 506,113 in 1997 therefore the average rate of domestic waste being disposed of was 0.238 kg/day in 1987 and 0.267 kg/day in 2007 per capita (UNDP/NISEPA, 2009). There is no remarkable increase in industrial waste within the two decades as the town did not experience much industrial growth, from 200tonnes in 1987 to 377tonnes in 2009. The percentage increase in institutional solid waste has increased. This was due largely to heavy presence of federal Agencies and location of Institutions in the town within the two decades. It increases from 487tonnes in 1987 to 1150tonnes in 2009 (Ogwueleka, 2009).

Table 2 presents the percentage distribution of solid waste in Minna from 1987 to 2009. Food waste remains the highest portion, closely followed by paper, and sanitary pads. The comparison of national waste statistics may not be too simple a task, due to the difference in compositional classifications and data gathering system, solid waste composition in Minna is quite similar to that in Kolkota, India (Hazra and Goel, 2009), but vary slightly from those in Phnom Penh city in Cambodia (Kum, et al, 2005). Food waste accounts for about 37% of the total waste streams and paper makes up 25%. Food and paper waste in Kolkota was about 39.6 and 25.5% of its total solid waste respectively (Hazra and Goel, 2009). There are about 68.0 and 48.7% of food and paper waste respectively in Phnom Penh almost doubled the percentage in Minna. The differences in solid waste composition can then have a serious impact on the techniques of solid waste management in different countries. Combustibility of a waste depends largely on its calorific values and varies substantially depending on the source and the period of the year. Therefore, incineration cannot be recommended generally for waste

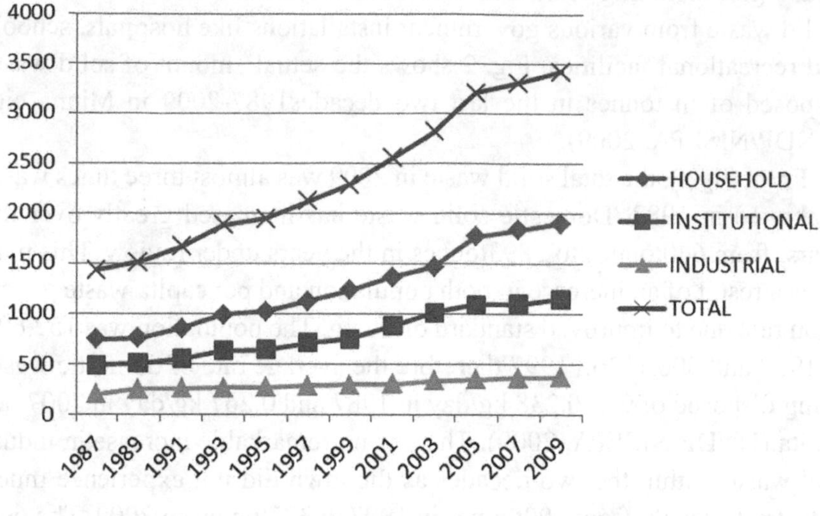

Figure 2. Waste Disposed of in Minna for the past two decades.

management unless the calorific properties of the waste are known and it has to be site specific (Ogwueleka, 2009).

Table 2 provides composition of the solid wastes produced in eight major locations in Minna. The main components are food residues, plastics, paper, glass bottles and metals. The table also shows that the plastic content is unusually high probably because it was wet, which increased the weight. Plastics mainly come from water and fruit juice bags and containers. Much of this material is in small pieces, mixed types; wet, dirty and hence recycling may be difficult. The results of the analysis also show that the generated waste in the city is largely organic matter that can be composted. The waste is also heterogeneous in composition comprising of both degradable and non-degradable materials, therefore sorting at site would have been an appropriate method for collection (Osman, 2009). The bulk of the non-degradable waste is potentially recyclable materials, while the degradable materials could be composted.

Table 1. Solid wastes composition in Minna between 1987 and 2007.

Composition of wastes (%)	Year				
	1987	1992	1997	2002	2007
Food/Organic wastes	36.45	36.61	40.58	42.10	42.58
Paper/Textiles/Leather	24.65	25.61	22.83	22.05	21.09
Plastics	6.02	6.91	5.64	5.93	5.91
Construction debris	5.60	5.60	8.65	8.69	8.75
Wood	3.22	2.64	2.64	2.69	3.01
Horticultural wastes	4.53	1.96	2.03	2.01	2.05
Metals (Ferrous and non-ferrous)	2.02	2.61	3.61	3.52	3.61
Sludge	0.09	0.07	1.20	1.45	1.49
Glass	1.02	1.04	1.15	1.14	1.11
Scrap tyres	0.44	0.61	0.94	0.94	0.87
Pampers/sanitary pads	5.26	6.58	7.59	7.59	7.84
Others	10.70	9.76	3.14	1.89	1.69

5.4 SOLID WASTES COLLECTION SYSTEM IN MINNA

Collection and transportation of waste is both labour and capital intensive. Waste transportation, including labour and machinery, accounts for between 70% and 80% of the total cost of solid waste management in Nigeria. (Imam et al, 2008). A shortage of waste collection vehicles in Minna is due to lack of funding and inadequate maintenance. Efficient collection depends on proper selection of vehicles; this needs to take account of road conditions, traffic density, availability of spare parts and servicing requirements. Waste collection service is available only in places where there are good roads. It was estimated that more than 35–40% of the population in the city is without regular or adequate collection service and the coverage, efficiency and frequency of collection vary substantially from one area to another (UNDP/NISEPA, 2009). Moreover, the transportation of waste

Table 2. Solid wastes composition in Minna between 1987 and 2007.

Waste Type (%)	Districts Name and characteristics							
	Chanchaga	Bosso	Tunga	Maikunkele	Kpakungu	Shango	Maitumbi	Tudun fulani
Food Remnants	51.14	56.40	53.41	49.61	53.22	49.61	49.36	49.26
Pampers/sanitary pads	6.41	5.69	5.66	5.39	6.21	4.36	4.84	4.16
Glass	5.91	4.33	6.41	6.29	5.62	5.16	5.44	7.01
Metals	7.92	6.41	6.54	6.23	6.43	6.44	6.41	6.33
Paper	2.46	5.61	3.44	4.49	2.46	1.46	1.23	3.99
Plastics	16.30	11.6	11.4	12.09	13.33	12.90	13.01	12.11
Wood	6.10	7.20	6.4	6.90	5.40	10.30	9.56	5.40
Others	0.76	2.76	6.74	9.00	7.33	9.77	10.15	11.74
Total	100	100	100	100	100	100	100	100

to the dump-site has not been properly managed. Wastes that are light in weight are clearly seen flying from the trucks during transport. This also contributes to the litter on streets. It seems that the collection service in the city is deteriorating, in many areas, the frequency of collection has dropped from once a day to once every three days and collection times are quite variable. As a result of this more and more households carry their waste to the nearby dumpsites popularly known as bolah inside waste-bags and thereby littering the street during transport. The waste bags can also be torn by scavenging animals and humans that search for something to eat searching for saleable materials. These activities usually scatter the waste that is ready for collection, and this makes the job of the collection crew even more difficult, as they have to shovel the scattered waste from the ground into the collection vehicle. This system leads to unacceptably low collection efficiency and too much waste is left on the streets of Minna.

The existing collection system should be replaced with a more efficient, but not more expensive system without delay. Waste scattered from the collection trucks during transportation is due to the lack of adequate cover during the trip. This can be alleviated by covering the waste during transport especially if the vehicle is travelling at more than 35 km/h. The use of compactor trucks for the transportation of waste as a means of enclosing waste during transport may also be considered, a compaction vehicle is designed for the waste volume reduction because on the average, such a vehicle will reduce the density of the waste to about 450–520 kg/m^3 from the initial density of 200–250 kg/m^3. However, the compactor trucks are very expensive and require high operating cost and also complex additional maintenance, but its usage will solve the problem of scattering of waste during transportation to the dump-sites (Sarwoko, et al, 2007).

5.5 WASTE DISPOSAL IN MINNA

Solid wastes from the different collection system in the various districts in Minna are transported to various dump-sites at the outskirts of the city. Piles of solid wastes are also found along roads, underneath bridges, in culverts and drainage channels and in other open spaces. This practice should not continue because it is not environmentally acceptable, and it makes environment unhealthy and unhygienic. The involvement of citizens in environmental sanitation is important, in Minna and of course in most Nigerian cities, every Saturday has been declared the environmental sanitation day. Most people now commit this day to clean their local environment. Civil servants now devote at least two hours a day in the week to cleaning their office premises. Through this measure, people are being made aware of the need to clean environment. The environmental sanitation day however causes problems, because people have no means of disposing the waste collected. Vehicle owners parked the wastes in their car and dump them along the major roads leading out of Minna. Though people participate in this cleaning up exercise, they should be educated on how to dispose of the waste properly.

It is still common in Minna to see people throw litter from cars or motor vehicles into streets while traveling and to see people in the parks leave litter on the streets even though rubbish bins are situated within walking distance. This is really a lack of responsibility and has a negative effect on the environment. People at times deliberately dump their waste into open channels thinking that it will be carried away by rainwater, not understanding the clogging and pollution problem this may cause. In the medium or higher income areas, the situation is a bit better people leave their waste inside plastic bags along the streets but these bags of waste become scattered by scavengers. This also shows that the general public has not been fully sensitized to participate actively in waste management issues. There is therefore a need for a greater public participation for better SWM in Minna city. The environmental issues can be included in the school curriculum so that the concept of waste management will grow with the students as they progress in life, this will build human resources for future generations.

5.6 FUTURE CHALLENGES

The changes in lifestyle, particularly in the urban areas, have led to more acute waste problems. The situation is further worsened in the sub-urban areas and in slum areas with additional problems of closely-packed housing and traffic, where air and water pollution are experienced.

Indiscriminate dumping in open places, access roads and watercourses are the problems that are widespread, which are human contribution to a public health problem. Thus, the challenges of sustainable development are population explosion lack of infrastructure and environmental pollution as causes and impacts. Wastes should be managed in such a way that our present and even coming generations will not be affected; this is because this environment is not inherited from our ancestors but just borrowed from our offspring (Turan et al, 2009). Scavenging activities should be discouraged. Soil cover should also be utilized, and the landfill be constructed. Recycling is still at zero level in Minna, nonetheless, with increasing environmental awareness; the government should start to promote waste recycling by drafting policies and offering support to private waste management companies. Waste minimization will remain to be one of the major future challenges; it therefore needs to be implemented more strictly. Currently, there is no limitation on the amount of solid waste that may be generated; minimization of residential solid waste will continue to be difficult until the regulation of Pay-As-You-Throw is fully implemented. Environmental protection campaigns should also be frequently launched, with the media always playing an important role. Biological treatment of organic solid wastes, such as composting and anaerobic digesting, has played an important role in many other countries. Food waste, for example, accounted for about 37% of the total solid waste in Minna but only 2.3% of it was recycle (Solomon, 2009). Non-toxic contaminated food waste should therefore be separated for biological treatment. This will reduce the energy consumption and cost needed for the incineration of food waste which is high in moisture content. Composted food waste can then be used for agricultural activities.

5.7 CONCLUSIONS AND RECOMMENDATIONS

It is widely accepted that solid waste management issues should be addressed from a system perspective by taking into account the technological, financial, institutional, legal, and socio-cultural factors to determine appropriate policies for the local surroundings (Salim, 2010). Rapid urbanization and population explosions has caused tremendous increase in solid waste generation in Minna. Appropriate storage systems at the sources of waste generation should be introduced into the waste management system in the city. Regulations regarding littering and improper disposal of solid waste should be formulated, and stiff penalties should be imposed on defaulters. Public awareness about the environment should be increased through environmental education so that the public participation in SWM will improve. A source reduction program should be encouraged and promoted because it is a way to address waste prevention. It is clear from government and general public actions that there is a need to have a clean environment. Many things still need to be done to ensure proper waste management. The campaign on environmental sanitation should be strengthened, it should be inculcated into daily life of every citizen, the number of people employed in environmental related jobs are so few that they cannot cope with the volume of waste generated, more people should be employed, cleaning up exercise should not be limited to when an important personality is visiting the city, it should be continuous and permanent. There should be only one major disposal area and should be operated as sanitary landfill site, this, though expensive, will be needed to eradicate littering the roadsides with rubbish.

REFERENCES

1. Adebayo WO, Bamisaye JA, Akintan OB, Ogunleye OS. 2006. Waste generation, disposal and management techniques in an urbanizing environment: A case Study of Ado-Ekiti, Nigeria. Research Journal of Applied Sciences 1(4), 63-66.
2. Anikwe MAN, Nwobodo KCA. 2002. Long term effect of municipal waste disposal on soil properties and productivity of sites used for urban agriculture in Abakaliki, Nigeria. Bioresource Technology 83(3), 241-250.
3. Ayo B, Ibrahim B, Mohammed RM. 2010. The practice and challenges of solid waste management in Damaturu, Yobe state, Nigeria. Journal of Environmental Protection 1, 384-388.
4. Chen X, Geng Y, Fujita T. 2010. An overview of municipal solid waste management in China. Waste Management 30(4), 716-724.
5. Hazra T, Goel S. 2009. Solid waste management in Kolkata, India: Practices and challenges. Waste Management 29(1), 470-478.
6. Imam A, Mohammed B, Wilson DC. Cheeseman, CR. 2008. Solid waste management in Abuja, Nigeria. Waste Management 28(2), 468- 472.

7. Izugbara CO, Umoh JO. 2004. Indigenous waste management practices among the Ngwa of Southeastern Nigeria: some lessons and policy implications. The Environmentalist 24(2), 87-92.
8. Kalu C, Modugu WW, Ubochi I. 2009. Evaluation of solid waste management policy in Benin metropolis, Edo State, Nigeria. African Scientist, 10(1), 117-125.
9. Kum V, Sharp A, Harnpornchai N. 2005. Improving the solid waste management in Phnom Penh city: a strategic approach. Waste Management 25(1), 101-109.
10. Manaf LA, Samah MAA, Zukki NIM. 2009. Municipal solid waste management in Malaysia: Practices and challenges. Waste Management 29(11), 2902-2906.
11. Ogwueleka TC 2009. Municipal solid waste characteristics and management in Nigeria. Iranian Journal of Environmental Health Science and Engineering 6(3), 173-180.
12. Osman NA. 2009. Comparison of old and new municipal solid waste management systems in Denizli, Turkey. Waste Management 29(1), 456-464.
13. Pokhrel D, Viraraghavan T. 2005. Municipal solid waste management in Nepal: practices and challenges. Waste Management 25, 555-562.
14. Salim CJ. 2010. Municipal solid waste management in Dar Es Salaam city, Tanzania. Waste Management 30, 1430-1432.
15. Sarwoko M, Agus PP, Alim FR, Ananto YS, Muhammad S. 2007. Priority improvement of solid waste management practice in java. Journal of Applied Science in Environmental Sanitation 2(1), 29-34.
16. Sha'Ato R, Aboho SY, Oketunde FO, Eneji IS, Unazi G, Agwa S. 2007. Survey of solid waste generation and composition in a rapidly growing urban area in Nigeria. Waste Management 27, 352- 358.
17. Solomon US. 2009. The state of solid waste management in Nigeria. Waste Management, 29, 2787-2790.
18. Turan NG, Atoru S, Akdemir A. Ergun ON. 2009. Municipal solid waste management strategies in Turkey. Waste Management 29(1), 465-469.
19. UNDP/NISEPA. 2009. Niger state framework for integrated sustainable waste management. Niger State Strategic Waste Management Framework

CHAPTER 6

Management of Municipal Solid Waste in One of the Galapagos Islands

MARCO RAGAZZI, RICCARDO CATELLANI, ELENA CRISTINA RADA, VINCENZO TORRETTA, AND XAVIER SALAZAR-VALENZUELA

6.1 INTRODUCTION

This paper relates to a study conducted on the island of Santa Cruz, Galápagos Archipelago, Ecuador. Given its exceptional and unique biodiversity, the archipelago enjoys special protection and laws. The total land area is 788,200 square kilometers, of which 96.7% consists of the National Park, and the remaining 3.3% is made up of urban and agricultural areas, located on the islands of San Cristobal, Santa Cruz, Isabela and Floreana. The entire province can be approximately divided into three strongly interconnected subsystems: a green park, a marine reserve and urban agglomerates [1].

In recent decades, the existences of a local population and increasing flow of tourists and heavy development have produced the first signs of the potential unsustainability of this fragile ecosystem. The national 2010 census reported a total population of 21,067 inhabitants, of whom 17,997 were in urban areas and the remaining 3070 were in rural areas. Population growth is one of the major emerging issues, resulting from tourism dynamics and the consequent increase in waste production [2–5].

Over the years, numerous measures have been proposed to address the problems related to the sustainable disposal of municipal solid waste (MSW). A curbside collection system has been recently established, together with a center for sorting and recycling [2].

The aim of this work is to analyze the criticalities arising from the need to preserve the ecosystems of this extraordinary island and the need to find an economically and technologically sustainable model of MSW management for the municipality of Santa Cruz, similarly to what is done in other low income countries [6–10]. A specific aim concerns the finding of replicable solutions for the management of MSW in tourist islands.

Our study focuses also on the new neighborhood under construction in Puerto Ayora in accordance with national legislation the Millennium Development Goals, national legislation, including the *Law of the Special Regime for the Conservation and Sustainable Development of the Province of Galapagos*, the Plan of Conservation and Sustainable Development developed by the former *Instituto Nacional Galápagos* (INGALA), and following the suggestions of the Pan American Health Organization [11–13].

6.2 METHODOLOGY

The present work has been developed through the following steps:

- Collection of literature data specifically available for the island of Santa Cruz
- Planning of field activity for the verification of the available data and their integration/updating
- Organisation and development of a 2 month stay of one of the authors for the planned field activities

- Critical analysis of the collected data, also for pointing out replicable experiences
- Elaboration of proposals for the improvement of the waste management

6.3 CHARACTERISTICS OF THE AREA OF STUDY

The island of Santa Cruz is the second largest island (986 km2) of the archipelago, with a maximum altitude of 864 meters above sea level and with the largest population in the province. The district of Santa Cruz includes the capital Puerto Ayora and the villages of Bellavista and Santa Rosa. In addition to the island of Baltra (site of the airport), the islands of Marchena, Pinta, Pinzon, Seymour and other smaller islets are under the jurisdiction of this district.

The population of the island has increased considerably: in 2010 the population reached 15,393 inhabitants, of whom 11,974 residents were in the urban area of Puerto Ayora, 2425 in Bellavista and 994 in Santa Rosa. The population density of Puerto Ayora is the largest of the entire island. In fact, 84.1% of the population is concentrated in this town distributed over an area of about 190 ha, with an average population density of 3.36 people per household [14]. The town of Puerto Ayora is divided into five sectors.

The municipality has decided to further expand the urbanized area, because the existing one is no longer sufficient to accommodate future population growth. For that reason the new neighborhood *El Mirador* was planned [13]. The project involves the construction of 1130 lots with a rectangular shape of 15 m × 20 m. The lots are grouped into sub districts (*manzanas*) with about 66 batches each, within which there is often an area of 1500 m^2 for recreational space. Public facilities are also provided in the center of the neighborhood to serve the community such as a sports field, a church, a medical clinic, and a children's playground.

The municipality has proposed certain types of homes that can be built. There are restrictions on the maximum number of buildings, density, and land use, the materials that can be used and on the heights of the fences. In *El Mirador* it is forbidden to open activities that could disturb the inhabitants such as bars, pubs, clubs, woodworking shops, machine shops, car washes, boat workshops and other potentially noisy or polluting activities [15]. The housing capacity of the new district includes 10,000 units [15], however using the current value of the inhabitants/dwelling (rounded up) a more appropriate value of just over 6000 inhabitants has been decided. Another 250 police officers and 100 soldiers, who will be transferred to the Mirador, need to be added, by moving two small barracks.

The collection of MSW of the few homes already inhabited is done "on call".

6.4 CURRENT WASTE MANAGEMENT

Throughout the district there is a service for the collection of MSW covering the whole of Puerto Ayora and the other two residential areas on the island. Where this service is not offered, people bury their own waste. The collection system (Figure 1) provides for the separation of different categories of material at source, leaving the user to divide the material into recyclable, organic and non-recyclable material (Figure 2). For hospitals and clinics a collection of "hospital" waste is provided.

Until 2009 there was not a technological landfill on the island and the waste was spread in a place 27 km away on the road to Baltra. To mitigate the pollution of this area, since 2011 there has been land reclamation and a real landfill is being built.

In the canton of Santa Cruz, MSW incineration (open burning) is carried out by very few citizens who are sensitive to the criticalities of this issue. This is a very positive aspect of the behavior of the local population.

6.4.1 MSW COLLECTION

The cleaning of streets, gardens, parks and other communal areas is carried out once a week on the busiest streets, every day in the commercial

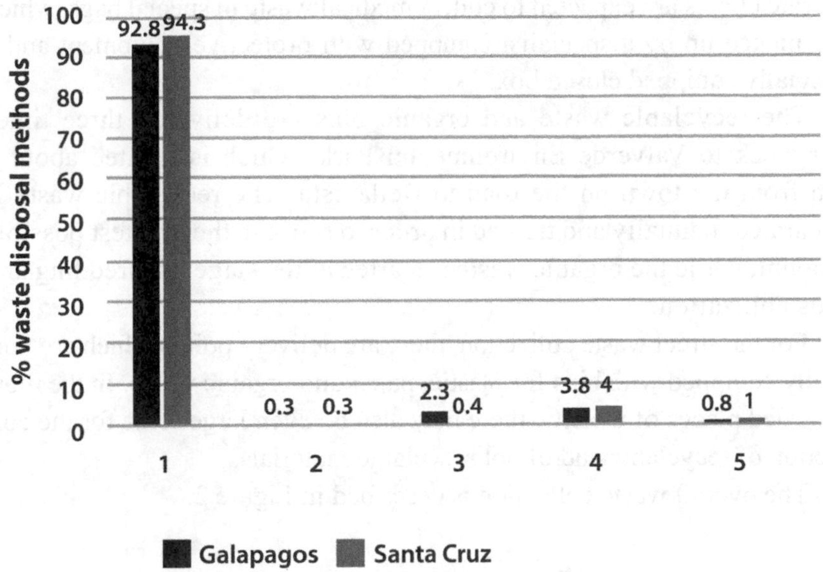

Figure 1. Comparison of municipal solid waste (MSW) disposal methods by the population of the Province of Galapagos, and those for the population of Santa Cruz [16] (1 = Municipal service; 2 = Abandonment on roads, rivers, soil; 3 = Incineration; 4 = Recycling/landfilling; 5 = Other).

areas located in the south of the city, twice a week in the residential areas, and twice a month in the peripheral area in the north of the city.

Solid waste is collected by three compactor trucks and two trucks with wooden boxes. Two teams are used to collect waste, performing on average two trips to the designated disposal/recovery, picking up approximately 10.5 tons of waste per day.

The waste collected as non-recyclable (residual MSW) and organic waste not suitable for composting are sent to the landfill site. The construction of the new landfill gave the possibility to manage them in a more organized way. Occasionally the scrap metal is taken away by private traders.

The curbside collection provides each user with three 70 L bins, for organic, recyclable and non-recyclable waste, respectively. Hospitals and private clinics are expected to collect medical waste in special bags, which are picked up by a specialist equipped with protective equipment and a specially equipped closed box.

The recyclable waste and organic bins are delivered three times per week to Valverde Environmental Park, which is located about 4 km from the town on the road to Bellavista. The recyclable waste is separated manually and treated in order to retrieve the greatest possible amount, while the organic waste is started at the stage of shredding and biostabilization.

For the street waste collection, there are delivery points which are normally equipped with bins for plastic, paper and organic waste. In the most crowded places of the city, there may also be even larger bins for the collection of recyclable and of not recyclable materials.

The overall waste collection is described in Figure 2.

6.4.1.1. ORGANIC WASTE

The organic waste, which is collected separately in green bins, is treated in a bio-stabilization plant and subsequently sent to a composting process. After the treatment, it is sieved and packed in 20 kg bags to be sold or stored. The municipality mainly uses this compost as fertilizer for flowerbeds and green spaces.

The average daily organic waste treated from 2009–2011, when this method developed quickly, was 1.98 tons/day. In Figure 3 the monthly values are reported.

6.4.1.2. RECYCLABLE WASTE

The recyclables collected are transferred to the Valverde recycling center where they are manually separated into various product fractions. The average daily quantity of separated waste treated is

1.47 tons. Details are reported in Table 1.

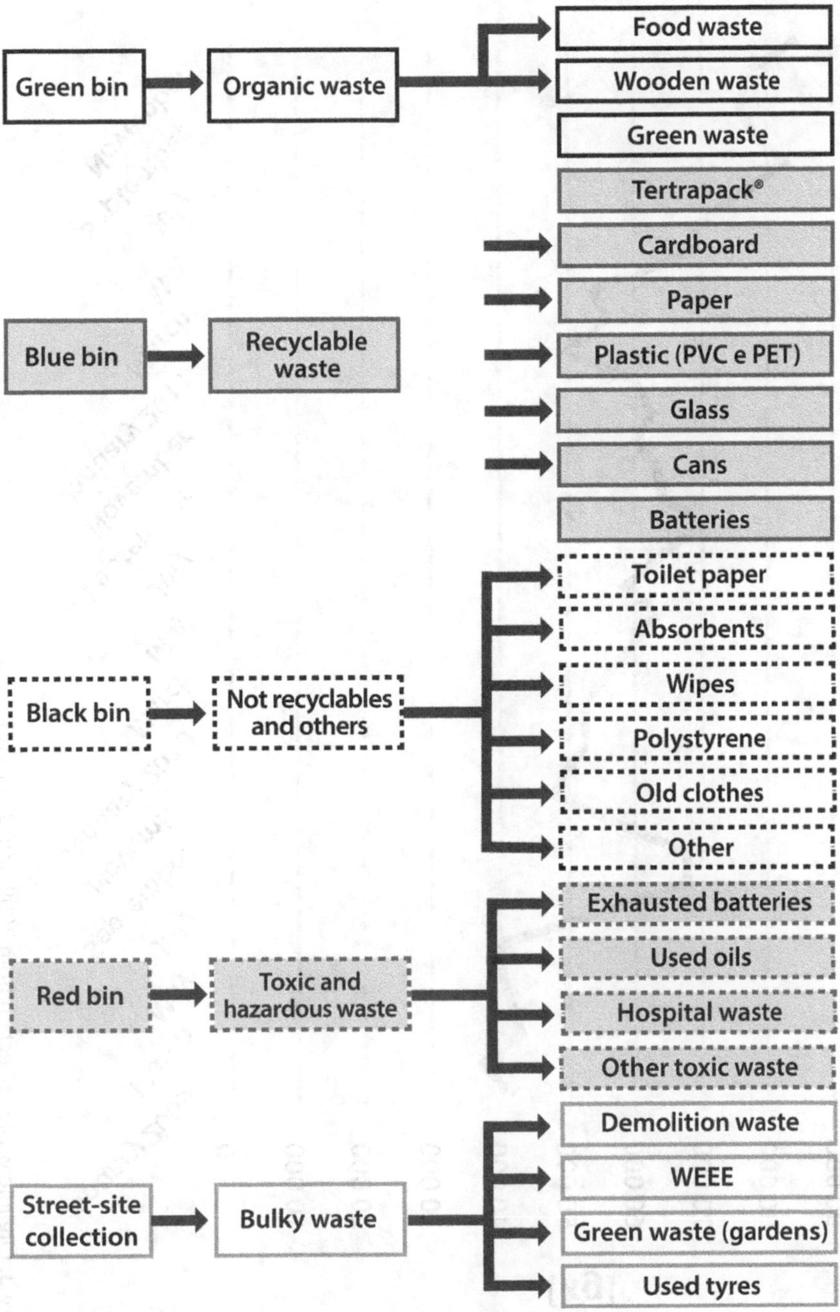

Figure 2. Waste collected with the current system.

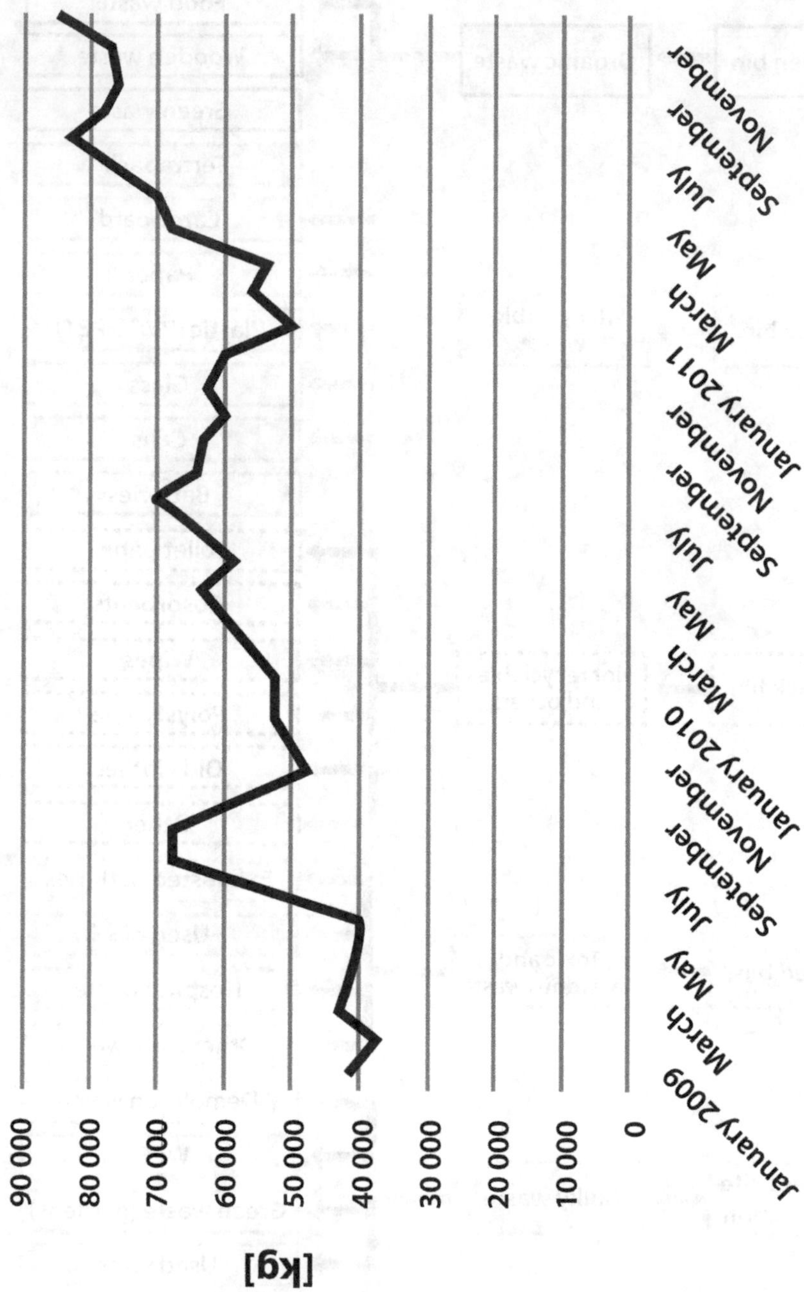

Figure 3. Organic waste collection from 2009 until 2011 [16].

6.4.1.3. RESIDUAL MSW

An average yield of 7.4 tons per day of residual MSW (that is non-recyclable waste and recyclable materials not source separated) is delivered to the landfill 27 km away [17]. In Figure 4 the monthly values are reported.

With this latest data, averaged over a time window of three years, it can be concluded that the organic fraction is about 18% of the treated waste, waste collected as recyclable is about 14%, while residual MSW accounts for approximately 68% of the total. By adding the two recoverable fractions, it is possible to reach 32% of recyclable material collected, which, in a context such as Ecuador, is certainly encouraging.

6.4.2 WASTE MANAGEMENT IN PUERTO AYORA

6.4.2.1 WASTE PRODUCED

In 2008, a study was carried out on the construction and safety of the landfill 27 km on behalf of the WWF. This study [18] showed that the most representative fractions of waste in MSW produced by the residents of Puerto Ayora and tourism are: kitchen organic waste, glass, cardboard and plastic bags (Table 2). Such significant data regarding packaging are not surprising, in fact, the archipelago is located very far from the mainland, and consequently the goods need more protection during transportation regardless of the means used. Plastic bags and diapers are two categories that commonly appear in the waste produced by residents.

The percentage of food waste is relevant in the present MSW (Table 3).

A projection of the amount of waste produced per capita until 2028 was estimated. The Table 4 shows the results of the study [18]. The role of organic fraction is expected to change in percentage; its absolute value (kg/inh/day) in Puerto Ayora is affected by the restriction to the import of fresh fruits from the continent: indeed fruit beverages must be prepared only from packed juices; more in general, any contamination with external seeds must be avoided. That reduces the per-capita generation of food waste compared to conventional cases.

Table 1. Recyclable materials, separated in the Valverde waste recycling center between 2009 and 2011 [16].

	Month	Cardboard (kg)	Glass (kg)	Plastic (kg)	Paper (kg)	Tetra-pack® (kg)	Cans (kg)	Packaging for Eggs (kg)	Batteries (kg)	Cement Bags (kg)	Total (kg)
2009	January	20,050	13,360	5121	2608	206	109	119.4	100	452	42,127
	February	18,389	9,560	3845	1453	267	636	343.2	0	1609.2	36,102
	March	14,055	14,135	5855	3709	437	226	91.4	0	689	39,197
	April	15,557	13,172	7738	3948	292	490	375.6	0	616	42,188
	May	18,295	15,336	5764	2823	270	430	232.8	0	596	43,747
	June	18,800	13,451	4920	2994	373	412	363	0	585	41,898
	July	22,379	14,721	4805	3050	526	559	368.6	0	936	47,344
	August	21,425	13,085	5347	2964	980	1025	932.6	0	623	46,380
	September	19,198	11,924	7169	4215	887	849	755.2	0	525	45,522
	October	15,663	9,144	4489	3258	442	561	338	0	1132	35,029
	November	19,932	10,620	4930	2838	818	402	0	0	713	40,253
	December	21,256	11,250	4951	4089	531	756	0	0	695	43,527

Table 1. Continued.

2010	January	21,099	14,345	4947	3774	885	1068	0	0	376	46,494
	February	20,254	11,177	5471	4562	873	612	0	0	427	43,376
	March	19,728	15,512	6314	6409	858	1099	0	0	204	50,124
	April	18,975	13,846	7659	4227	1004	512	0	0	1160	47,384
	May	19,193	12,594	5987	3676	861	496	0	0	455	43,262
	June	17,638	11,505	6127	3845	1204	384	0	1209	1245	431,577
	July	18,959	13,107	5523	3795	959	187	0	0	728	43,257
	August	17,414	12,995	5416	5001	1037	225	0	0	636	42,724
	September	17,917	14,120	5255	3224	733	341	0	0	668	42,259
	October	17,440	13,170	4563	3142	807	376	0	0	1372	40,871
	November	19,868	14,137	5160	3103	1708	277	0	0	480	44,732
	December	24,578	15,132	4349	3619	764	346	0	0	584	49,373

Table 1. Continued.

2011	January	18,757	13,790	4737	6581	0	610	0	0	1286	45,761
	February	15,973	13,864	7194	4175	0	346	0	0	0	41,552
	March	17,621	17,041	6345	4388	0	709	0	0	0	46,103
	April	20,604	15,665	6264	4666	0	348	0	0	166	47,714
	May	19,577	15,156	6253	3614	0	287	0	0	338	45,225
	June	19,648	14,167	5404	4073	0	678	0	0	408	44,377
	July	19,686	14,660	7331	4723	0	814	0	0	611	47,825
	August	18,620	12,741	6190	3862	0	490	0	368	713	42,984
	September	19,165	12,084	5564	3112	0	450	0	0	364	40,739
	October	18,513	13,578	4880	3074	0	490	0	0	792	41,327
	November	18,707	15,755	6027	4210	0	385	0	0	960	46,044
	December	21,398	17,556	7209	4850	0	460	0	0	1100	52,573

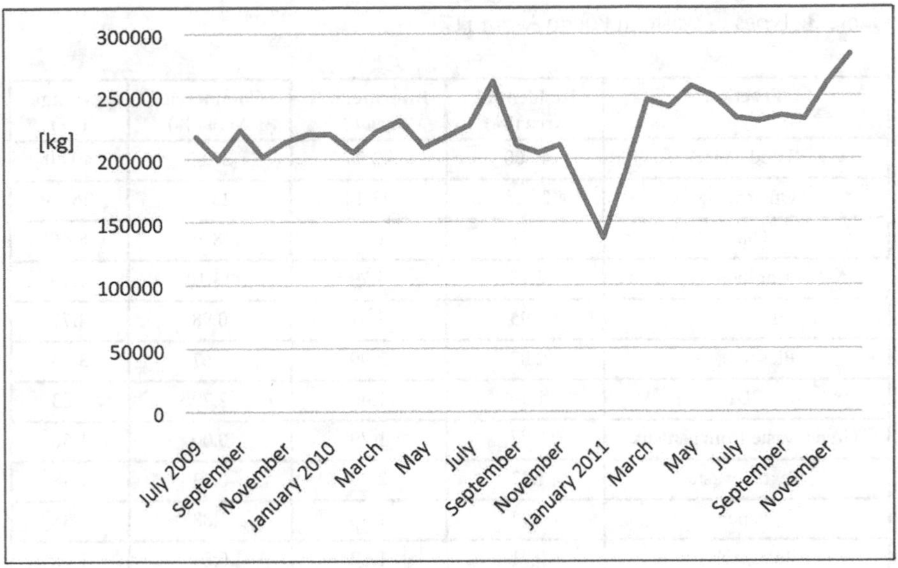

Figure 4. Residual MSW collection from 2009 until 2011 [16].

6.4.2.2. MSW MANAGEMENT

Plastic bags are also used for waste before depositing them in recycle bins. In the stream of source separated recyclable waste and residual MSW, plastic bags are not a problem, because in the former they are separated manually and in the latter they are taken directly to the landfill. However, organic waste needs to be carefully removed to avoid contamination of the final product.

The waste from the municipal slaughterhouse, the public market and the fish cleaning and packing firm are dumped in the landfill. Waste from demolition and scrap from the building industry have the same fate.

An effort must be made for improving the management of tires, used batteries and a part of green waste, but this can be seen as a demonstration

Table 3. Types of waste in Puerto Ayora [17].

Fraction	Residential Area (%)	Intermediate Area (%)	Commercial Area (%)	Average (%)
Food waste	49.00	43.77	27.83	40.20
Other waste	26.08	27.14	26.36	26.53
Glass	5.31	1.69	18.39	8.46
Cardboard	2.98	4.79	13.10	6.96
Diapers	4.95	8.37	0.98	4.77
Plastic bags	2.07	2.99	4.67	3.24
PET	1.30	1.40	2.79	1.83
Green waste from gardens	2.72	1.79	0.00	1.50
Textile waste	1.43	2.79	0.23	1.48
Paper	0.93	1.34	1.88	1.38
Tetrapack®	0.71	1.69	0.83	1.08
High density plastics	0.97	1.00	1.06	1.01
Scrap metal	1.17	0.64	0.83	0.88
Batteries	0.00	0.00	0.90	0.30
Shoes	0.26	0.50	0.00	0.25
Trays	0.00	0.00	0,15	0.05
Polystyrene foam	0.06	0.10	0.00	0.05
Copper wires	0.06	0.00	0.00	0.02

Table 4. Waste generation: future expectation [18].

Flux	MSW Production expected in 2028 (kg/inh/day)
Food waste	0.211
Residual MSW	0.315
Recyclable waste	0.155
Green waste from gardens	0.154
Market and slaughterhouse waste	0.022

that the optimization of waste management is obtainable through a pathway of years.

Paper, cardboard, scrap metal, paper packaging bags of cement and plastic are separated manually and safely by operators equipped with individual protection systems. After being compressed and packaged, the waste is sent to the mainland and sold as shown in Table 5.

Recently, the separation of waste from electrical and electronic equipment (WEEE) has also begun, which is packed and sent to the mainland, and which costs around US$ 300/ton for disposal. Cells and batteries are separated from the recyclable waste stream, and at the moment are stored under a shed awaiting a suitable place for disposal [18].

For a certain period, a small incinerator was operating for the disposal of hospital waste. A discussion was opened about the acceptability of combustion in this context. An alternative approach based on the autoclave principle was presented a few years ago as an alternative.

6.5 POTENTIAL IMPROVEMENTS OF THE WASTE MANAGEMENT

The above described situation refers to a 3 year period when the management of waste in the island showed significant improvements. In this chapter, potential improvements to be considered for additional modifications are discussed.

6.5.1 CO-COMPOSTING OF ORGANIC AND GREEN WASTE FROM GARDENS

An interesting alternative for the disposal of sewage sludge is to co-compost it with organic waste from the MSW from households, restaurants, hotels, markets and garden waste.

Numerous experiments have been carried out [19,20] on co-composting two or more streams of organic matter. The combination of sewage or faecal sludge/organic fraction of MSW is the most commonly studied, because the combined treatment of this waste would result in considerable

Table 5. Gain from sale of recyclable material [18].

Material	Gain (US$/ton)
Glass	60
Plastic	150
Paper	120
Scrap metal	260

savings compared to separate management. In addition, these waste streams can be processed very well together because they are complementary. Organic waste is particularly rich in organic carbon [21], while faecal sludge has a good content of nitrogen and moisture, the fundamental parameters for the successful outcome of the end product.

The characteristics of the output product are strongly affected by the quality of the incoming products and the techniques used for composting. Wastes from pruning, grass cuttings and leaves are the most valuable fractions in the co-composting process [22]. The high C/N ratio makes it advisable to compost waste containing a high concentration of nitrogen as sewage sludge.

Normally you try to get at least one sewage sludge sedimentation or dry fixed bed dehumidification. This pre-treatment is essential to maintain a mixing ratio of faecal sludge vs. organic matter of MSW as close as possible to 0.25–0.5, however up to 0.1 [23] is also possible in the case of untreated sludge.

Faecal sludge can be co-composted with any type of biodegradable material. Drying beds could help its management. It is advisable to leave the mature compost for 90 days in order to have a better product and it is hygienic to turn the piles over every 10 days. Given the variability of the water content of the input material, it is necessary to change the ratios of mixture so that the moisture of the material to be composted is maintained in the range of optimal values equal to 50%–60% [19].

The municipality should also have a shredder to reduce the volume of incoming waste, green waste, especially from gardens. After shredding, mixing could take place in the bioreactor which is already owned by the Valverde recycling center.

This technique avoids the need to burn plant cuttings. Once this process has been started, we suggest cleaning the area where, to date, this waste (and often also non organic waste) is burned. Larger logs of wood, not for grinding, can be recovered for other uses (i.e., boat building, furniture or other objects).

One of the problems encountered in the treated and sieved material, is the pieces of plastic bags that are used to contain the organic household waste. A good initiative by the Galapagos National Park (GNP) and the Galapagos Government Council might be to ban plastic shopping bags and use envelopes made of biodegradable or recycled paper. Further progress could be made by banning imports of disposable plastic (such as dishes and cutlery).

The economic sustainability of a composting plant also depends on the revenues from the sales of compost. The current process did not guarantee a high quality product, but the above suggested practices should ensure an increase in the fertilizing properties of this product.

In recent years, politicians in Ecuador have worked hard to develop a more responsible model of organic farming towards small farmers and the environment. The integral agrarian reform introduces the concept of food sovereignty, and encourages people to produce food mainly for their own consumption, avoiding the use of chemical agents and methods of cultivation which over time can damage the soil. Increasing the quality and production of the organic fertilizer produced in this process appears to be consistent with the guidelines of the central state.

According to studies [17], in the future 0.387 kg/inh/day of compostable material will be produced, while the production of dried sludge in the neighborhood El Mirador is estimated to be equal to 0.043 kg/inh/day. Considering a 2036 population of 19,516 inhabitants in Puerto Ayora and 6021 for the new expansion, an estimated production of compostable waste amounts to 3607.2 tons/year whilst for treated sludge a 94.6 tons/year is expected. The small quantity of the latter fraction of the total allows for easy dilution during composting.

Summing up, the above mentioned measures aimed to co-composting could help in decreasing the impact of the management of biodegradable fractions.

6.5.2 WASTE BATTERIES

Ecuador has no specialized facilities for the recovery and treatment of waste batteries. The product fractions separated in the sorting center also include exhausted batteries. For now, these are stored in bins under a shed in the Valverde recycling center, pending the setting up of a place where they can be permanently placed. Alternatively, one of the four tanks intended for the ashes of hospital waste has also been considered, because the tanks are considered as sufficiently safe.

By placing two rows of bins full of batteries, would be able to form inside the tank two levels of drums, to be sealed to prevent leakage. Each tank is 3 m × 2 m in plan and 1.5 m in height, and could hold 48 cans filled with two rows of 24, plus another row of bins with a maximum height of 40 cm.

To increase the security, a row could be divided with the other planks of wood recovered from the collection and use as a barrier and stabilizer for the drums of batteries.

It would be good practice to encourage the use of rechargeable AA batteries and a mini stylus in order to decrease the production of exhausted batteries. An effective information campaign could also be addressed to tourists to take home the exhausted batteries.

These measures could help in decreasing the amount of heavy metals not correctly managed.

6.5.3. WASTE CONTAINING ASBESTOS

In a recent census, of the 4270 buildings examined in Puero Ayora, the roofs of 3549 buildings were covered as follows:

- 1303 concrete (slabs, cement);

- 1210 asbestos cement (Eternit®, Eurolit®);
- 1032 zinc;
- 4 other materials.

About one third are asbestos cement coverings, known under the trade names of Eternit® or Eurolit®. In the case of the asbestos roofs, the risks depend on the probability of releasing asbestos fibers into the air and/or in the soil, which in turn is linked to the state of preservation of the product itself. Where in the surrounding parts of the asbestos roof, fiber dispersion occurs, the removal, encapsulation or constructing above the roof, covering it may be required. Asbestos should be removed by a highly skilled labor force in order to ensure maximum security for the reclamation and disposal of roofing elements containing asbestos. To proceed with the necessary checks on the state of conservation of the asbestos cement roofing and possible safe disposal, appropriate technical training is recommended.

In Ecuador, a moratorium is proposed to prohibit imports from the mainland of this type of product. Moreover, the island does not have a proper system for the removal and final disposal of this material.

6.5.4 MSW MANAGEMENT IN THE NEW URBANIZED AREA IN EL MIRADOR

The new urbanization will be served by the same curbside collection system that already exists. It is recommended that an interior space in each lot is acquired, which is easily accessible from the outside area, to place the bins for the collection of recyclable, non-recyclable and organic materials.

One of the problems most often encountered in the streets of Puerto Ayora is that green waste is left in the streets. It is therefore necessary to obtain a clearance of at least 8 m^2, protected with a fence or a gate, where the population is allowed to dispose of the garden and green waste separately with those in small demolitions and tires. These small ecological islands should be installed in each sub-lot (in total 21) or preferably in the free areas in the middle or in those small free areas that are occasionally found near the sidewalks.

In this way, the entire neighborhood would be fully serviced by facilities with a catchment area with a radius of approximately 120 m and a population of a little more than 230 people. Both the radius and number of people are reasonably low and make it possible to improve the collection of these MSW fractions, as long as the storage areas are emptied three times a week as currently happens for curbside collection in the town.

6.5.5 OTHER RECOMMENDATIONS

It is important that decisions are taken with regard to the reduction of waste. For example, a campaign may be useful to sensitize employees to the reduction of printed paper or at least to reuse the printed paper for rough copies of documents.

Many of the privately owned gardens are full of unused and discarded objects. The municipalities could introduce a law for the protection of public decorum to facilitate the disposal of this waste. For example, a monthly market for citizens could be held to exchange and barter certain goods.

The cleaning and maintenance of roads could be probably not sufficient to guarantee an optimal service for the city. Incentives for the creation of groups of volunteers to help the municipal workers might help give citizens a strong sense of collaboration. These organizations could promote artistic recycling courses to involve more and more citizens in the fight against waste and the uncontrolled production of MSW.

6.6. CONCLUSIONS

The study enabled us to build a picture of the waste management in the study area, highlighting the positive experiences and suggesting some improvements.

Concerning the positive experiences in the area, it must be pointed out that the collection of mixed packaging coupled with a manual selection plant should be taken into account for a replication in similar realities where the tourist fluxes increase this king of waste fractions.

A more careful waste management would transform the operations that are normally considered to be problematic, ensuring a better quality of the organic waste and recyclables separation by a more careful collection. In addition, actions have been outlined for the appropriate disposal of hazardous and non-hazardous municipal waste, aimed at safeguarding the environment in line with the principles of sustainable development.

In our opinion, this study can be a useful approach to address similar problems everywhere, with particular attention to the low income countries and oceanic islands.

REFERENCES

1. Charles Darwin Research Station (CDRS); Parque Nacional Galapagos (PNG). Galápagos Report 2009–2010. Puerto Ayora, Ecuador. Available online: http://www.galapagospark.org/oneimage.php?page=ciencia (accessed on 3 September 2014).
2. Gobierno Autónomo Descentralizado Municipal de Santa Cruz. Diagnóstico sistema asentamientos humanos (Diagnosis Human Settlement System). Puerto Ayora, Ecuador. Available online: http://www.gobiernogalapagos.gob.ec/wp-content/uploads/downloads/2013/08/PDOT-Santa-Cruz -2012_2_segundo1.pdf (accessed on 3 September 2014).
3. Parque Nacional Galapagos (PNG). Control and eradication of introduced animals. Puerto Ayora, Ecuador. Available online: http://www.galapagospark.org/nophprg.php?page=parque_nacional_ introducidas_animales (accessed on 3 September 2014).
4. Parque Nacional Galapagos (PNG). Control and eradication of introduced plants. Puerto Ayora, Ecuador. Available online: http://www.galapagospark.org/nophprg.php?page=parque_nacional_introducidas_plantas (accessed on 3 September 2014).
5. Watkins, G.; Cruz, F. Galapagos at Risk—A Socioeconomic Analysis; Charles Darwin Foundation: Puerto Ayora, Ecuador, 2007.
6. Collivignarelli, C.; de Felice, V.; Di Bella, V.; Sorlini, S.; Torretta, V.; Vaccari, M. Assessment of sanitary infrastructures and polluting loads in Pojuca river (Brazil). Water Pract. Technol. 2012, 7. doi:10.2166/wpt.2012.044.
7. Torretta, V.; Conti, F.; Leonardi, M.; Ruggieri, G. Energy recovery from sludges and sustainable development: A Tanzanian case study. Sustainability 2012, 4, 2661–2672.
8. Vaccari, M.; Torretta, V.; Collivignarelli, C. Effect of improving environmental sustainability in developing countries by upgrading solid waste management techniques: A case study. Sustainability 2012, 4, 2852–2861.

9. Rada, E.C.; Ragazzi, M.; Fedrizzi, P. WEB-GIS oriented system viability for municipal solid waste selective collection optimization in developed and transient economies. Waste Manag. 2013, 33, 785–992.
10. Rada, E.C.; Zatelli, C.; Mattolin P. Municipal solid waste selective collection and tourism. WIT Trans. Ecol. Environ. 2014, 180, 187–197.
11. Environmental Ministry of Ecuador. Texto Unificado de Legislación Ambiental (Environmental regulation). Available online: Available online: http://www.google.it/url?sa=t&rct=j&q=&esrc =s&source=web&cd=1&ved=0CCIQFjAA&url=http%3A%2F%2Fwww.miliarium.com%2Fpaginas%2Fleyes%2Finternacional%2FEcuador%2FGeneral%2FTextoUnificado%2FLibroVI-Anexo1.doc&ei=119AUPijM8r-4QSiZw&usg=AFQjCNEJzwhh8zfH3q_YsRiQUZ (accessed on 3 September 2014).
12. Organization of the Petroleum Exporting Countries (OPEC). OPEC Annual Statistical Bulletin 2012. Available online: http://www.opec.org/opec_web/static_files_project/media/downloads/publications/ASB2012.pdf (accessed on 3 September 2014).
13. United Nations (UN). The Millenium Development Goals. Available online: http://www.un.org/millenniumgoals/bkgd.shtml (accessed on 3 September 2014).
14. Instituto Nacional de Estatisticas y Censos (INEC). Censo de poblacion y vivienda 2011 (Census of Population and Housing 2011). Quito, Ecuador. Available online: http://www.ecuadorencifras.gob.ec/ (accessed on 3 September 2014).
15. Gobierno Autónomo Descentralizado Municipal de Santa Cruz. Proyecto de "Urbanizacion El Mirador" en terrenos y el Gobierno Municipal de Santa Cruz (Project of urbanization "El Mirador" in the Municipal Government of Santa Cruz); Santa Cruz Municipality: Puerto Ayora, Ecuador, 2010.
16. Instituto Nacional de Estatisticas y Censos (INEC). Encuesta de condiciones de vida en Galapagos 2009–2010; Report on Life Conditions in Galapagos; Quito, Ecuador, ye. Available online: http://www.ecuadorencifras.gob.ec/ (accessed on 3 September 2014).
17. De la Torre, F. Relleno Sanitario para la Isla Santa Cruz, Provincia de Galápagos (Characterization of solid waste in the Islands Santa Cruz, San Cristobal and Isabela, Galapagos); Municipality of Santa Cruz, Office of Environmental Managment, Puerto Ayora, Ecuador, 2009.
18. Plan de manejo de desechos para las Islas Galapagos (Waste Management Plan for Galapagos Islands); World Wildlife Fund (WWF): Quito, Ecuador, 2010.
19. Cofie, O.; Kone, D.; Rothenberger, S.; Moser, D.; Zubruegg, C. Co-composting of faecal sludge and organic solid waste for agriculture: Process dynamics. Water Res. 2009, 43, 4665–4675.
20. Rada, E.C.; Ragazzi, M.; Villotti, S.; Torretta, V. Sewage sludge drying by energy recovery from OFMSW composting: Preliminary feasibility evaluation. Waste Manag. 2014, 34, 859–866.
21. Andreottola, G.; Ragazzi, M.; Foladori, P.; Villa, R.; Langone, M.; Rada E.C. The unit integrated approach for OFMSW. UPB Sci. Bull. 2012, 74, 19–26.
22. Bikovens, O.; Dizhbite, T.; Telysheva, G. Characterization of humic substrances formed during co-composting of grass and wood wastes with animal grease. Environ. Technol. 2012, 33, 1427–1433.

23. Eawag/Sandec—Department of Water and Sanitation for Developing Countries. Faecal Sludge Management—Sandec Training Tool, Module 5; Doulaye Konè & Sylvie Peter: Dubendorf, Switzerland, 2008.

PART III

STRATEGIES

CHAPTER 7

Application of Bioremediation on Solid Waste Management: A Review

TIWARI GARIMA AND S. P. SINGH

7.1 INTRODUCTION

'Earth' has rich wealth of natural resources such as land, forests, wildlife, soil, air, water, wind, animals and plants. The race began when humans started living a stable life rather than a nomadic life. Civilization's use, over use, and misuse has led to depletion of various natural resources to an extent that today half of our natural resources are either depleted or at the edge of depletion [1].

And due to civilization, urbanization and industrialization, large amounts of wastes are generated, which are dumped into the environment annually. Approximately 6×10^6 chemical compounds have been synthesized, with 1,000 new chemicals being synthesized annually. Almost 60,000 to 95,000 chemicals are in commercial use. According to third world network reports, more than one billion pounds (450 million kilograms) of toxins are released globally in air and water. The contaminants causing ecological problems leading to imbalance in nature is of global concern. At the international level the researchers of the world are trying to overcome the depletion of natural resources by several means; however very little attention is given to their words and most of the world continues to use them without caring the adverse consequences. The dumping of hazardous waste into the environment (like rubber, plastics, agricultural waste, and industrial waste) is harmful to living creatures.

Solid-waste management is a major challenge in urban areas throughout the world. Without an effective and efficient solid-waste management program, the waste generated from various human activities can result in health hazards and have a negative impact on the environment. Continuous and uncontrolled discharge of industrial and urban wastes into the environmental sink has become an issue of major global concern [2,3]. The industrial and anthropogenic activities have also led to the contamination of agricultural lands which results in the loss of biodiversity. Although the use of pesticides and herbicides increases the productivity of crop but also increase the contamination in the soil, water and air [4].

Bioremediation is not only a process of removing the pollutant from the environment but also it is an eco-friendly and more effective process [5]. The pollutants can be removed or detoxified from the soil and water by the use of microorganism, known as bioremediation [6,7]. The purpose of bioremediation is to make the environment free from pollution with the help of environmental friendly microbes. Bioremediation broadly can be divided in two categories: i.e In-situ bioremediation and ex-situ bioremediation.

This study reviewed the salient features of methods of bioremediation, its limitations and recent developments in solid waste management through bioremediation.

In situ bioremediation provides the treatment at contaminated sites avoiding excavation and transport of contaminants, which means there is no need to excavate the water or contaminated soil for remediation. There is a biological treatment of cleaning the hazardous substances on the surface. Here the use of oxygen and nutrient is at the contaminated site in the form of aqueous solution in which bacteria grow and help to degrade the organic matter. It can be used for soil and groundwater.

Generally, this technique includes conditions such as the infiltration of water containing nutrients and oxygen or other electron acceptors for groundwater treatment [8]. Most often, in situ bioremediation is applied to the degradation of contaminants in saturated soils and groundwater. It is a superior method to cleaning contaminated environments since it is cheaper and uses harmless microbial organisms to degrade the chemicals. Chemotaxis is important to the study of in-situ bioremediation because microbial organisms with chemotactic abilities can move into an area containing contaminants. So by enhancing the cells' chemotactic abilities, in-situ bioremediation will become a safer method in degrading harmful compounds. This in-situ bioremediation is further sub divided into the following categories:

7.1.1 BIOVENTING

This is a technique to degrade any aerobically degradable compound. In bioventing the oxygen and nutrient (like nitrogen and phosphorus) is injected to the contaminated site [9]. The distribution of these nutrient and oxygen in soil is dependent on soil texture. In bioventing enough oxygen is provided through low air flow rate for microbes. Bioventing is nothing but pumping of air into contaminated soil above the water table through a well which sucked the air. Bioventing is more effective if the water table is deep below the surface and the area has a high temperature. It is mainly used for the removal of gasoline, oil, petroleum etc. The rate removal of these substances is varied from one site to another site. This is because of the difference in soil texture and different composition of hydrocarbons (Figure 1 and Table 1).

7.1.2 BIOSPARGING

In biosparging air is injected below the ground water under pressure to increase the concentration of oxygen. The oxygen is injected for microbial degradation of pollutant. Biosparging increases the aerobic

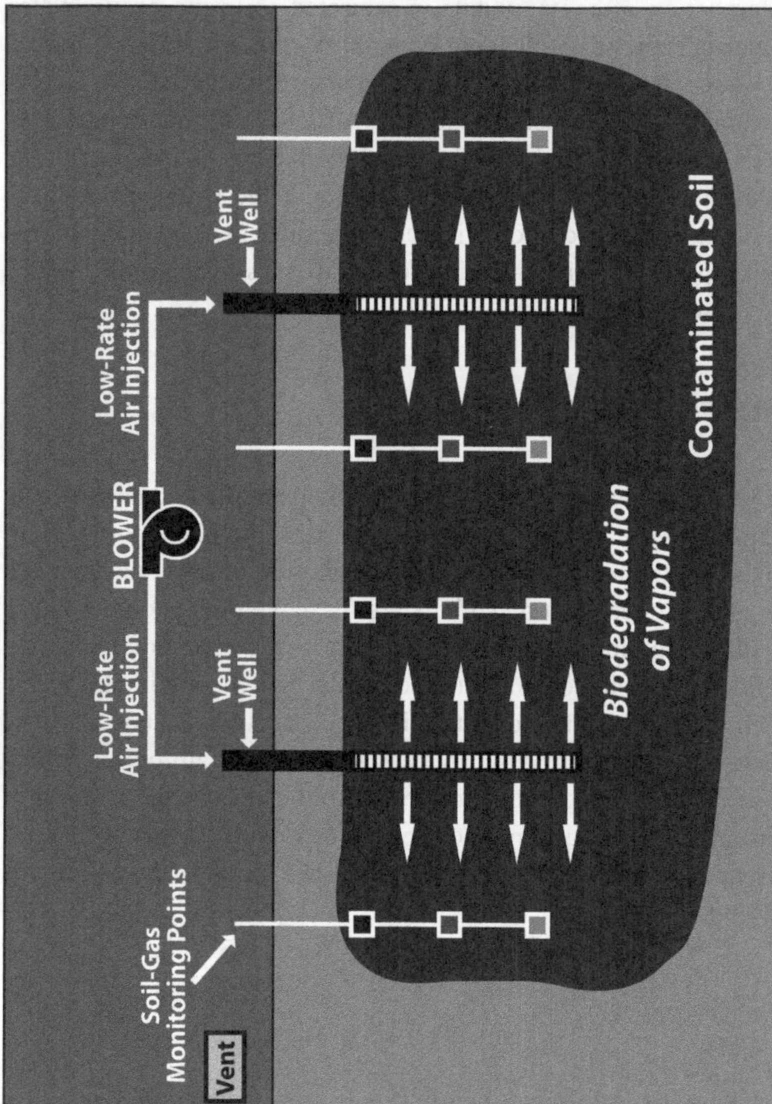

Figure 1. Schematic diagram of a typical bioventing system.

Table 1. A thesis on Bioventing Degradation Rates of Petroleum Hydrocarbons and Determination of Scale-up Factors by Alamgir Akhtar Khan, The University of Guelph.

Sr no	Organization	Accidents occurred at	Key note	References
1	USEPA	United State	58100 UST leak have been identified	[28]
2	United Kingdom	Hazardous Installations Directorate	409 dangerous occurrences for 2011/12	[29]
3	Canadian federal contaminated sites inventory	Canada	46.6 % of total contamination concerns soil, with 52.1 % of soil contamination due to petroleum	[30]
4	England	--	17 % of all serious contamination incidents in 2007 were related to fires, spills and leaks of hydrocarbons	[31]

degradation and volatilization [10]. There must be control of pressure while injecting the oxygen at the contaminated site to prevent the transfer of volatile matter into the atmosphere. The cost can be reduced by reducing the the diameter of injection point. Before injecting the oxygen, soil texture and permeability should be known. This technology was applied to a known source of gasoline contamination in order to quantify the extent of remediation achieved in terms of both mass removed and reduction in mass discharge into groundwater. Biosparging is effective in reducing petroleum products at underground storage tank (UST) sites. Biosparging is most often used at sites with mid-weight petroleum products (e.g., diesel fuel, jet fuel); lighter petroleum products (e.g., gasoline) tend to volatilize readily and to be removed more rapidly using air sparging. Heavier products (e.g., lubricating oils) generally take longer to biodegrade than the lighter products, but biosparging can still be used at these sites. Even after that there are some disadvantages.

Advantages	Disadvantages
It is readily available and easy to install	It can be used in environmental where air sparging is uniform, permeable soil, unconfined aquifer etc
Treatment time is short and very minimal disturbance to the operation site	There is no field and laboratory data to support design consideration

7.1.3 BIOAGUMENTATION

Microorganisms having specific metabolic capability are introduced to the contaminated site for enhancing the degradation of waste. At sites where soil and groundwater are contaminated with chlorinated ethenes, such as tetrachloroethylene and trichloroethylene, bioaugmentation is used to ensure that the in situ microorganisms can completely degrade these contaminants to ethylene and chloride, which are non-toxic. Monitoring of this system is difficult (Figure 2).

7.2 EX-SITU BIOREMEDIATION

The treatments are not given at site. In ex situ, the contaminated soil is excavated to treat it at another place. This can be further sub divided into following categories:

7.2.1 BIOPILING

This is a hybrid form of composting and land farming. The basic biopile system includes a treatment bed, an aeration system, an irrigation/nutrient system and a leachate collection system. For proper degradation there should be control of moisture, heat, nutrients, oxygen, and pH. The irrigation system is buried under the soil and provides air and nutrient through vacuum. To prevent the run off the soil is covered with plastic, due to which evaporation and volatilization is also prevented the solar

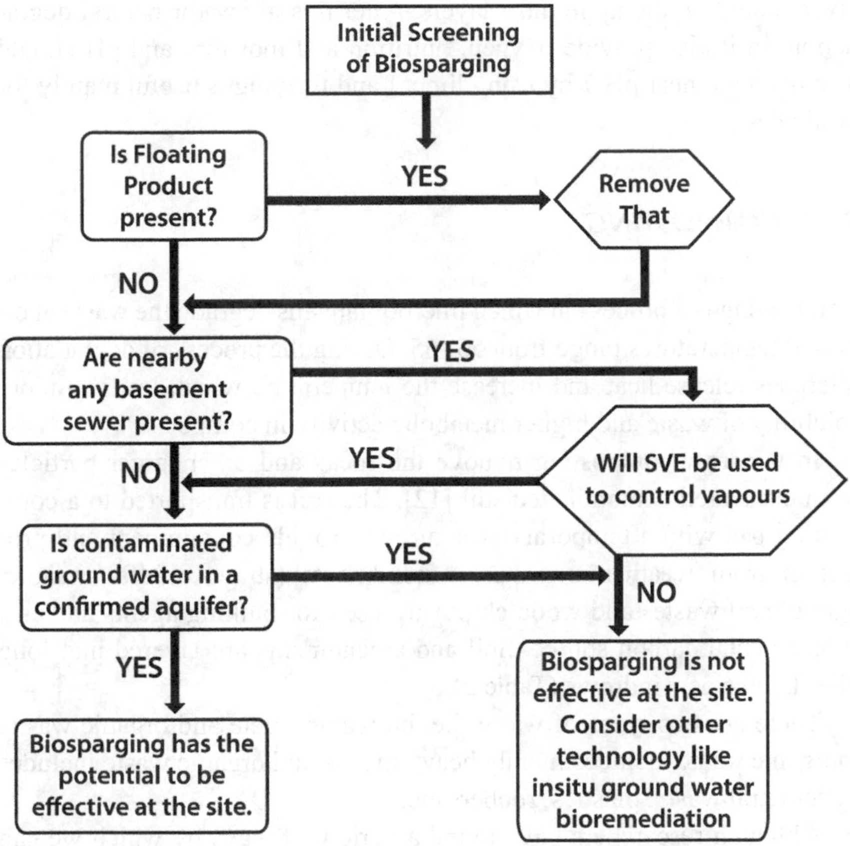

Figure 2. Biosparging process flow chart.

heating is promoted. Biopile treatment takes 20 to 3 months to complete the procedure [11].

7.2.2 LANDFORMING

In land forming a sandwich layer of excavated soil is made between a clean soil and a clay and concrete. The clean soil at bottom and concrete layer should be the uppermost layers. After this allow for natural degradation. In it also provide oxygen, nutrition and moisture and pH should also maintain near pH 7 by using lime. Land forming is useful mainly for pesticides.

7.2.3 COMPOSTING

Composting is a process in which microorganisms degrade the waste at elevated temperatures range from 55- 65. During the process of degradation microbes release heat and increase the temperature which leads to more solubility of waste and higher metabolic activity in composts.

In windrow composting remove the rocks and other larger particles from excavated contaminated soil [12]. The soil is transported to a composting pad with a temporary structure to provide containment and protection from weather extremes. Amendments (straw, alfalfa, manure, agricultural wastes and wood chips) are used for bulking agents and as a supplemental carbon source. Soil and amendments are layered into long piles known as windrows (Table 2).

There are two types of waste: i.e. inorganic waste and organic waste. Inorganic waste includes mainly heavy metals and organic waste includes agricultural waste, plastics, rubbers etc.

Although researchers have found a variety of ways by which we can degrade the solid waste, bioremediation is also making its leap to tackle the problem of heavy metals associated with different categories of waste with the help of microorganism (Figure 3).

7.3 BIOREMEDIATION OF HEAVY METALS

The atomic weight and density of heavy metal is high as compared to other elements. There is more than 20 heavy metals, but only a few of them such as Cadmium (Cd), Copper (Cu), Argon (Ar), Silver (Ag), Chromium (Cr), Zinc (Zn), Lead (Pb), Uranium (Ur), Nickel (Ni) etc. are considered, due to their toxicity. The contaminations of soil through heavy metals has become a major problem among all other environmental problems. These

Table 2. Work done on different methods of bioremediation and its applications.

Technique	Type	Application	Special technique	Removal	References
In Situ	Bioventing	Useful for hydrocarbons removal from contaminated site	---	Petroleum	[11]
			A blower or s compressor is connected to air supply well and soil gas monitoring well	Petroleum	[15]
			Air is injected at low flow rate for 15 month	Non-fuel hydrocarbonlike-acetone	[32]
	Biospaging	Indigenous micro organism are useful in presence of metals	Most efficient Non Invasive	Hydrocarbon	[33]
	Bioaugmentation	Useful for soluble chemical	Naturally attenuated process, treat Soil and water. Remove toxic material	For the treatment of waste water	[34]
			bench-scale batch andcontinuous flowactivated sludgereactors	For waste water treatment	[35]
			Use nitrogen as a essential component	For waste water treatment	[36]
				For removal of Chlorinated organic	[37]
	Land farming	Aerobic process and useful for organic material followed by irrigation and tailing	Inexpensive , self-heating Cost efficient, Simple,		[38]

Table 2. Continued.

		Anaerobic process converts organic solids to humus	Low cost Rapid reaction rate, Inexpensive, self-heating		[38]
	Composting		Using White rot fungi	Lignin degradation	[39]
			Use of cellulase, xylanase, manganese peroxidase, lignin peroxidase and Laccase during composting	For the degradation of lignin, cellulose and hemicellulose.	[40]
			During composition maintain moisture 75% and pressure under 0.6 bar	Compostingof organic materials from municipal solid waste	[41]

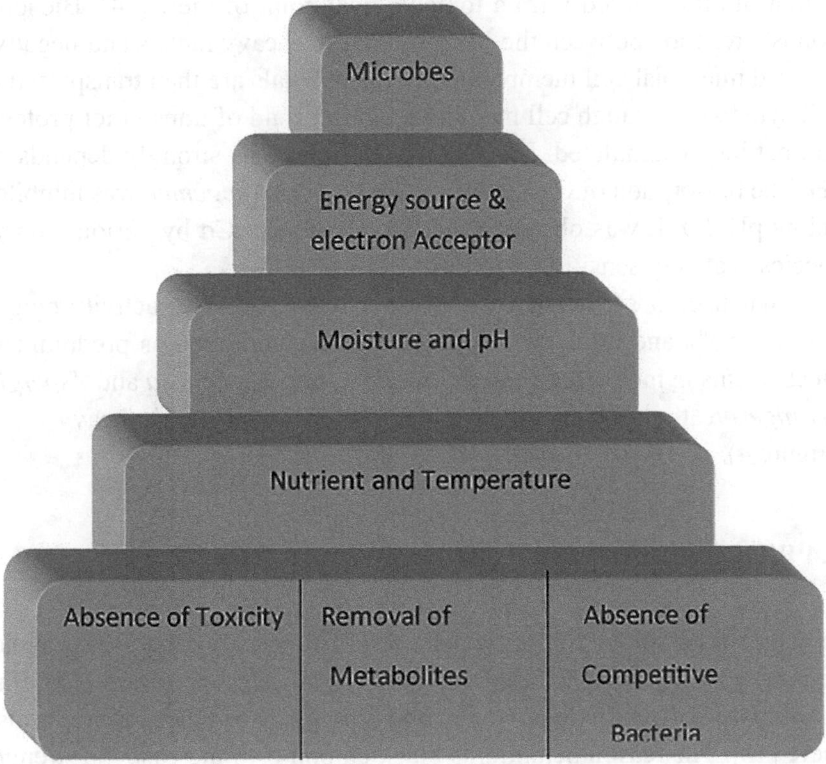

Figure 3. Requirements of bioremediation.

heavy metals contaminate not only the soil but also ground water through leaching. The removal of heavy metal is very important due to their potential of entering into the food chain causing adverse effect to human beings which accumulate in the body. These metals can also be removed by the use of various biological agents like yeast, fungi, bacteria, and algae etc. which act as bio sorbent for sequestering the metals. It can sequester dissolved metal ions out of dilute complex solutions very quickly, which is more effective and efficient. Hence it is an ideal candidate for the treatment of high volume and low concentration complex wastewaters [13]. The property of the microorganism to accumulate/sequester the metal is first of all observed form a toxicological point of view [14]. Biosorption is a reaction between the positive charged heavy metals and negative charged microbial cell membrane, in which metals are then transported to cell cytoplasm through cell membrane with the aid of transporter proteins and get bio accumulated. Biosorption of metal ions strongly depends on pH. The biosorption of Cr, Zn, Ni and Pb by *p. chrysogenum* was inhibited below pH 3.0. It was observed that biosorption of Cd by various fungal species is at very sensitive pH (Table 3).

It has been observed that Cd^{2+} Cr^{6+} and Zn^{2+} removal activity ranged between 85% and 60%, with intracellular accumulation as predominant mechanisms in most of the cases. *Pseudomonas aeruginosa* and *Aspergillus nigerare* the species which remove almost every toxic heavy metal (Figure 4).

7.4 BIOREMEDIATION OF RUBBER WASTE

In solid waste, about 12% is constituted of rubber. A rubber can neither degrade easily nor be recycled due to its physical composition [15]. Tire is composed of synthetic polymers and high grade of black carbon is also there [16]. The reason behind this black carbon is to increase the strength of that rubber tire [17]. A major environmental problem arises due to rubber, because on burning it gives a large number of toxic fumes along with carbon monoxide [18]. Even after that the use of rubber is increasing day by day, of which maximum rubber comes from vehicles i.e. 65% [19]. Its toxic chemical composition like zinc oxides inhibit the growth of sulfur oxidizing and other naturally occurring bacteria, which leads to slow natural degradation of rubber [20]. So for degradation of rubber first of all remove the toxic component of rubber through fungi like *Recinicium bicolour*. After that this rubber can be devulcanized by sulfur reducing or oxidizing bacteria like *Pyrococcus furiosus* & *Thiobacillus ferroxidans*. These devulcanized rubbers can be recycled [21]. The calorific value of rubber is same as coal, that is of about 3.3 x 10^4 KJ/kg [22]. So control combustion of rubber can be a best waste management [15] and the heat can be used for energy generation.

Table 3. Table showing the name of microbial species & Removal elements.

S. no.	Name of the species	Removal of elements	Reference
1	*Bascillus species*	Cd, Cu, Zn	[42,43]
2	*Cellulosmicrobiumcellulans*	Cr	[44]
3	*Pseudomonas aeruginosa*	Cd, Pb, Fe, Cu, U, Ra, Ni, Ag	[45,46]
4	*Aspergillus fumigates*	Ur	[13]
5	*Aspergillusniger*	Cd, Zn, Th, Ur, Ag, Cu	[43,47]
6	*Beta vulgaris*	Cd, Ni,Cr, Hg,	[43]
7	*Micrococusroseus*	Cd	[48]
8	*Escherichia coli*	Zn and V	[49]
9	*Oedogoniumrivulare*	Cr, Ni, Zn, Fe, Mn Cu, Pb, Cd and Co	[44]
10	*TrichodermaViride, And HumicolaInsolens*	Hg	[50]

7.5 BIOREMEDIATION OF AGRICULTURAL WASTE

Each year, humans, livestock, and crops produce approximately 38 billion metric tons of organic waste worldwide. Disposal and environmental friendly management of these wastes has become a global priority. Therefore, much attention has been paid in recent years to develop low-input and efficient technologies to convert such nutrient-rich organic wastes into value-added products for sustainable land practices. These can be managed through vermicomposting. Vermicomposting is a joint action between the earthworms and microorganism. Here microorganisms help in the degradation of organic matter and earthworms drive the process and conditioning to the substrate and altering the biological activity [23,24].

Several epigeic earthworms, e.g., *Eisenia fetida* (Savigny), *Perionyx excavatus* (Perrier), *Perionyx sansibaricus* (Perrier), and *Eudrilus*

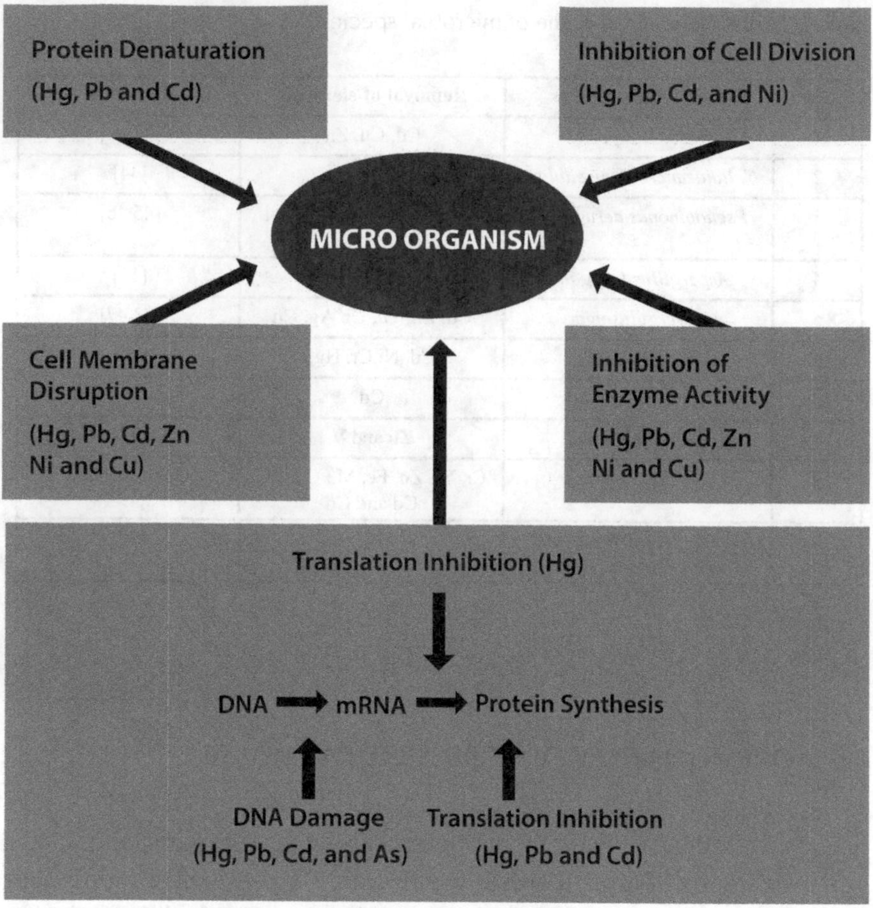

Figure 4. Toxicity of heavy metals also affects the microbial population.

eugeniae have been identified as detritus feeder and can be used potentially to minimize the anthropogenic waste from different sources [25], whereas agricultural by products like animal dung, crop residue etc. are good sources of nutrient for plants. In India, according to conservative estimation approximately 600 to 700 million tons of agricultural waste is available. This huge quantity of waste can be converted to boifertilizer by vermicomposting. Vermicomposting often results in mass reduction, shorter time for processing, and high levels of humus with reduced phytotoxicity in ready material [26]. A variety of combinations of crop residues and cattle manure were used in vermicomposting trials to obtain a value-added product, i.e., vermicomposting, at the end, the higher concentrations of plant nutrients in end products indicate a potential for using agriculture wastes in sustainable crop production [27].

7.5.1 DEGRADATION OF XENOBIOTIC COMPOUNDS.

Xenobiotics are organic in nature and many of the xenobiotic compounds released into the environment accumulate because they are only degraded very slowly and in some cases so slowly as to render them effectively permanent (Figure 5).

A short summary of some cardinal issues of significance for all Xenobiotics has been given bellow.

- The degradation of xenobiotic compounds depends upon microbial activity. Some example includes degradation of parathion
- The degradation pathway of xenobiotic compound should be examined when single substrate is available there.
- In the absence of oxygen there should be an alternative electron accepter, such as nitrate, sulphate, selenate, carbonate etc.
- There are no microbes or group of microbes that degrade all compounds. So there should be a group of organism, metabolically versatile that is applicable for the degradation of large amounts of compound.

The degradation of xenobiotic compound through white rot fungi can take place with certain enzymes. It has been reported that the degradation of TNT can be accomplished by non-ligninolytic strains of *P. chrysosporium* (Table 4).

7.6 LIMITATION OF BIOREMEDIATION

Some common environmental limitations to biodegradation are related to hazardous chemical wastes which possess high waste concentrations

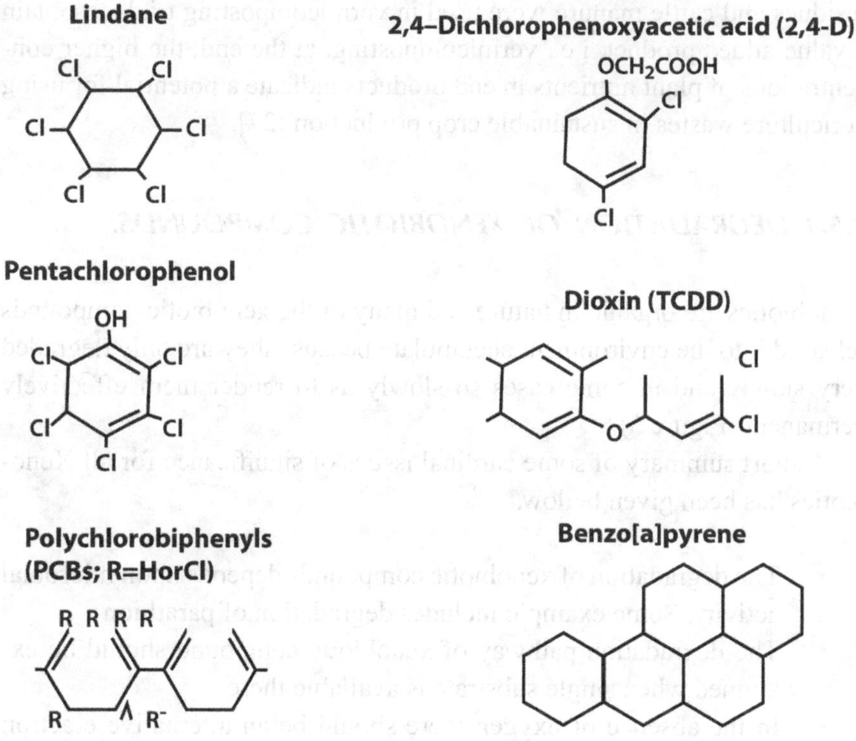

Figure 5. The structures of some herbicides, pesticides, and environmental contaminants.

and toxicity, because sometimes this toxicity either inhibits the growth of microorganism or sometimes kill them. For proper growth of microorganism it requires favorable pH condition and sufficient amount of mineral nutrients and also requires temperatures on which maximum microbes can survive i.e. 20°C to 30°C. Once environmental conditions are corrected, the ubiquitous distribution of microorganisms, in most cases, allows for a spontaneous enrichment of the appropriate microorganisms. In the great majority of cases, an inoculation with specific microorganisms is neither necessary nor useful. Besides all these, some other factors also effect the bioremediation such as solubility of waste, nature and chemical composition of waste, oxidation-reduction potential of waste and microbial

Table 4. Table showing the name of microbial species & Removal elements.

Xenobiotic compound	Microbes	Reference
Endosulfan compounds	*Mycobacterium sp.*	[51]
Endosulphate compounds	*Arthrobacter sp.*	[52]
Vinylchloride	*Dehalococcoides sp.*	[53]
Napthalene	*Pseudomonas putida*	[50]
Pyrene	*MycobacteriumPYR-1*	[54]
	Sphingomonaspaucimobilis	[52]
PCB	*RhodococcusRHA1*	[22]
Benzene	*Dechloromonas sp.*	[26]

interaction with this. Hence researchers should search genetically different types of microbes which can also work on slightly adverse conditions. Therefore, bioremediation is still considered as a developing technology to regulate the day to day environmental problems faced by humans residing in an area.

7.7 CONCLUSION

Although researchers have found a variety of ways by which we can degrade solid waste, but bioremediation is also making its leap to tackle the problem associated with different categories of waste with the help of microorganism. We conclude that less work has been done on rubber waste degradation, calling for more attention to rubber waste.

REFERENCES

1. Gosavi K, Sammut J, Gifford S, Jankowski J (2004) Macroalgalbiomonitors of trace metal contamination in acid sulfate soil aquaculture ponds. Sci Total Environ 324: 25-39.

2. Gupta R, Mohapatra H (2003) Microbial biomass: an economical alternative for removal of heavy metals from waste water. Indian J ExpBiol 41: 945-966.
3. Strong PJ, Burgess JE (2008) Treatment methods for wine-related ad distillery wastewaters: a review. Bioremediation Journal 12: 70-87.
4. Kumari R, Kaur I, Bhatnagar AK (2013) Enhancing soil health and productivity of Lycopersiconesculentum Mill. UsingSargassumjohnstoniiSetchell and Gardner as a soil conditioner and fertilizer. J ApplPhycol 25:1225-1235.
5. Singh SN, Tripathi RD (2007) Environmental bioremediation technologies, Springer-Verlag Berlin Heidelberg.
6. Talley J (2005) Introduction of recalcitrant compounds. In W. Jaferey& L. Talley (Eds. Bioremediation of recalcitrant compounds. Boca Raton: CRC.
7. Wasi S, Jeelani G, Ahmad M (2008) Biochemical characterization of a multiple heavy metal, pesticides and phenol resistant Pseudomonas fluorescens strain. Chemosphere 71: 1348-1355.
8. VidaliM (2001) Bioremediation An overview, Pure Appl. Chem73: 1163-1172.
9. Rockne K, Reddy K (2003) Bioremediation of Contaminated Sites, University of Illinois at Chicago.
10. Lambert JM, Yang T, Thomson NR, Barker JF (2009) Pulsed biosparging of a residual fuel source emplaced at CFB borden, Inter. J. Soil, Sedi. Water.
11. Niu GL, Zhang JJ, Zhao S, Liu H, Boon N, et al. (2009) Bioaugmentation of a 4-chloronitrobenzene contaminated soil with Pseudomonas putida ZWL73. Environ Pollut 157: 763-771.
12. Blanca AL, Angus JB, Katarina S, Joe LR, Nicholas JR (2007) The influence of different temperature programmes on the bioremediation of polycyclic aromatic hydrocarbons (PAHs) in a coal-tar contaminated soil by in-vessel composting. Journal of Hazardous Materials, 14:340-347.
13. Wang J, Chen C (2006) Biosorption of heavy metals by Saccharomyces cerevisiae: a review. BiotechnolAdv 24: 427-451.
14. Volesky B (1990) Removal and recovery of heavy metals by biosorption. In: Volesky B, editor. Biosorption of heavy metals. Florida: CRC press Pp 8-43.
15. Conesa JA, MartÃn-GullÃ³n I, Font R, Jauhiainen J (2004) Complete study of the pyrolysis and gasification of scrap tires in a pilot plant reactor. Environ SciTechnol 38: 3189-3194.
16. Tsuchii A, Tokiwa Y(2006) Microbial degradation of the natural rubber in tire tread compound by a strain of Nocardia. J. PolymerEnviron. 14:403-409.
17. Larsen MB, Schultz L, Glarborg P, Skaarup-Jensen L, Dam-Johansen K, et al. (2006)Devolatisation characteristics of large particles of tire rubber under combustion conditions, Fuel 85: 1335-1345
18. AdhikariB, De D, MaitiSD(2000) Reclamation and recycling of waste rubber. Prog. Polymer Sci. 25:909-948.
19. Holst O, Stenberg B, Christiansson M (1998) Biotechnological possibilities for waste tyre-rubber treatment. Biodegradation 9: 301-310.
20. Zabaniotou AA, Stavropoulos G(2003) Pyrolysis of used automobile tires and residual char utilization, J. Anal. Appl. Pyrolysis70:711-722.
21. Keri S, Bethan S, Adam GH (2008) Tire Rubber Recycling and Bioremediation, Bioremediation Journal, 12:1-11

22. Rajan V (2005)Devulcanization of NR based latex products for tire applications: Comparative investigation of different devulcanization agents in terms of efficiency. PhD Thesis, University of Twente, Enschede, the Netherlands.
23. Dominguez J (2004) State-of-the art and new perspectives on vermicomposting research. In: Earthworm Ecology, C.A. Edwards, Baca Raton, FL: CRC Press.
24. Suthar S (2007) Nutrient changes and biodynamics of epigeic earthworm Perionyxexcavatus (Perrier) during recycling of some agriculture wastes. BioresourTechnol 98: 1608-1614.
25. Garg P, Gupta A, Satya S (2006) Vermicomposting of different types of waste using Eiseniafoetida: a comparative study. BioresourTechnol 97: 391-395.
26. Lorimor J, Fulhage C, Zhang R, Funk T, Sheffield R, et al. (2001) Manure management strategies/technologies. White Paper on Animal Agriculture and the Environment for NationalCenter for Manure and Animal Waste Management. Ames, IA: Midwest Plan Service
27. SurindraSuthar(2009) Bioremediation of Agricultural Wastes through Vermicomposting, Bioremediation Journal, 13: 21-28
28. Eyvazi M.J, Zytner RG (2010) A Correlation to Estimate the Bioventing Rate Constant. Bioremediation Journal, 13:141-153.
29. United Kingdom Health & Safety Executive (2012) Offshore injury, ill health and incident statistics 2011/2012. HID statistics report HSR 2012-1.
30. Office of Auditor General of Canada (2012) Chapter 3-Federal contaminated sitesand their Impacts.
31. United Kingdom Annual Report of Incidents (2007) Food standards agency.
32. Sayles GD, Leeson A,Trizinsky MA, Rotstein P (1997)Field Test of Nonfuel Hydrocarbon Bioventing In Clayey-Sand Soil. Bioremediation Journal. Taylor& Francis Group, London, UK 1:123-133
33. Bouwer EJ, Zehnder AJ (1993) Bioremediation of organic compounds--putting microbial metabolism to work. Trends Biotechnol 11: 360-367.
34. Stephenson D, Stephenson T (1992) Bioaugmentation for enhancing biological wastewater treatment. BiotechnolAdv 10: 549-559.
35. Qasim SR, Stinehelfer ML (1982) Effect of a Bacterial Culture Product on Biological Kinetics. Journal Water Pollution Control Federation Pp. 255
36. Bouchez T, Patureau D, Dabert P, Juretschko S, DorÃ© J, et al. (2000) Ecological study of a bioaugmentation failure. Environ Microbiol 2: 179-190.
37. Boon N, De Gelder L, Lievens H, Siciliano SD, Top EM, et al. (2002) Bioaugmenting bioreactors for the continuous removal of 3-chloroaniline by a slow release approach. Environ SciTechnol 36: 4698-4704.
38. Antizar-Ladislao B, Katerina S, Angus JB, Nicholas JR (2008) Microbial communitystructure changes during bioremediation of PAHs in an aged coal-tar contaminated soil by in-vessel composting, International Biodeterioration& Biodegradation, 61: 357-364.
39. Tuomela M, Vikman M, Hatakka A, Itavaara M(2000) Biodegradation of lignin in a compost environment: a review, Bioresource Technology 72: 169-183
40. Sjostrom E (1993) Wood Chemistry, Fundamentals and Applications. (2ndedn) Gulf Professional Publishing Houston, Texas.

41. Abdelhadi M, Omar A, Mohammed M (2013) Effect of initial moisture content on the in-vessel composting under air pressure of organic fraction of municipal solid waste in Morocco. Iranian Journal of Environmental Health Science & Engineering.
42. Philip L, Iyengar L, Venkobacher L (2000) Site of interaction of copper on Bacillus polymyxa. Water Air Soil Pollution 119: 11-21.
43. Rajendran P, Muthukrishnan J, Gunasekaran P (2003) Microbes in heavy metal remediation. Indian J ExpBiol 41: 935-944.
44. Chatterjee S, Gupta D, Roy P, Chatterjee NC, Saha P, Dutta,[2011] S. Study of a lead tolerant yeast strain BUSCY1. Afr J Microbiol Res 5:5362-5372.
45. Jayashree R, Nithya SE, Rajesh PP, Krishnaraju M (2012) Biodegradation capability of bacterial species isolated from oil contaminated soil. J Academia Indust Res 1:140-143.
46. Pattus F, Abdallah M (2000)Siderophores and iron-transport in microorganisms: Review. J Chin Chem Soc. 47:1-20.
47. Guibal E, Roulph C, Le Cloirec P (1995) Infrared spectroscopic study of uranyl-biosorption by fungal biomass and materials of biological origin. Environ Sci-Technol 29: 2496-2503.
48. Motesharezadeh B (2008) Study of possibility of increasing phytoremadiation efficiency of heavy metal-contaminated soil by biological factors.Ph.D. Thesis, University College of agriculture and natural resource, Tehran University.
49. Grass G, Wong MD, Rosen BP, Smith RL, Rensing C (2002) ZupT is a Zn(II) uptake system in Escherichia coli. J Bacteriol 184: 864-866.
50. Muhammad MJ, Ikram-ul-Haq, Farrukh S (2007)Biosorption of Mercury from Industrial Effluent by Fungal Consortia, Bioremediation Journal 11: 149-153
51. Sutherland TD, Horne I, Russell RJ, Oakeshott JG (2002) Gene cloning and molecular characterization of a two-enzyme system catalyzing the oxidative detoxification of beta-endosulfan. Appl Environ Microbiol 68: 6237-6245.
52. Weir KM, Sutherland TD, Horne I, Russell RJ, Oakeshott JG (2006) A single monooxygenase, ese, is involved in the metabolism of the organochloridesendosulfan and endosulfate in an Arthrobacter sp. Appl Environ Microbiol 72: 3524-3530.
53. He J, Ritalahti KM, Yang KL, Koenigsberg SS, LÃ¶ffler FE (2003) Detoxification of vinyl chloride to ethene coupled to growth of an anaerobic bacterium. Nature 424: 62-65.
54. Kanaly RA, Bartha R, Watanabe K, Harayama S(2000) Rapid mineralization of Benzo[a]pyerene by a microbial consortium growing on diesel fuel. Appl. Environ. Microbiol 66:4205-4211.

CHAPTER 8

Community Engagement and Environmental Life Cycle Assessment of Kaikōura's Biosolid Reuse Options

JAMES E. MCDEVITT, ELISABETH (LISA) R. LANGER, AND ALAN C. LECKIE

8.1 INTRODUCTION

Kaikōura is a small relatively remote community on the South Island of New Zealand with a strong commitment to protecting the environment and working towards sustainability for their community and visitors. The Kaikōura District Council's wastewater treatment plant serves a permanent population of approximately 3,500 and a tourist population of up to one million visitors per year. Sewage sludge was dredged from Kaikōura's oxidation ponds six years ago (the first dredging for 25 years) and about 1,500 tons have been stockpiled. The community and council are tasked with deciding the most appropriate way of managing the stabilized sewage sludge (biosolids) before the current stockpiling consent runs out in 2016. Biosolids are carbon-rich and contain valuable nutrients and can reduce dependence on artificial fertilizers [1,2]. However, biosolids can contain a range of micro-contaminants such as heavy metals, pathogens and pharmaceuticals and personal care products [3]. Therefore, this biosolid waste stream is both a potential source of soil improvement and pollution.

The study reported here is an additional study allied to the publicly funded research program called the Biowastes Programme. The Biowastes Programme has been developed to better understand the environmental risks and benefits that can arise from applying biowastes to land with a multi-disciplinary team from research institutes, universities, Iwi (largest social unit in Māori culture) and local businesses. The main biowaste that the program has been focusing on is biosolids, and the principal focus is case-study research combining biophysical science with social research involving community, rūnanga (representative Māori assembly) and government regulators. Environmental and biophysical research has been undertaken to characterize the Kaikōura biosolids to provide stakeholders with information for their decision-making. Following an initial hui (a social assembly in a Māori community) and interviews with key stakeholders, a second community engagement hui was held with key stakeholders to select reuse options for the stockpiled biosolids and provide insights into community views on contaminants; this was held at the Takahanga marae (a marae is a sacred place that serves religious and social purposes in Polynesian societies) in Kaikōura, February 2011. After presentation of science results, a facilitated workshop session was held to enable key stakeholders and regulators to discuss a number of feasible options for their biosolids. A total of 19 options were presented to the community; further stabilization (six options); land application (five options); rehabilitation of land (four options); and resource recovery (four options). Participants were asked to discuss the environmental positives and negatives, social and cultural positives and negatives, economics and feasibility of each of the options. Community stakeholders at this hui identified a number of reuse options that were realistically available to the community, these were:

- Open air composting: a facility at the Innovative Waste Kaikōura site will be constructed. This will compost a biosolids garden green waste mixture; the resulting compost will be made available for sale to the community.

- Vermi-composting: where a vermi-composting facility is made and the biosolids and garden green waste are passed through the digestive tract of worms in containers. This produces relatively high quality compost and byproducts (i.e. worm juice and worms) that will be sold.
- Mixture with Biochar: the biosolids are mixed with a biochar material made from forest waste residues using slow-pyrolysis technology and made available for land application.
- Farm Application: the biosolids are applied directly to pastoral farmland, albeit ensuring that the biosolids remain outside of the human food chain.
- Forest Application: using existing forestry machinery, biosolids are applied to Radiata pine (*Pinus radiata L.*) plantation forests. The biosolids can be included as a soil amendment in the establishment of the stand.
- Land Rehabilitation: as part of the reclamation of marginal land biosolids are applied as soil amendment in order to promote vigorous plant growth of native tree species.

Making decisions on complex systems is particularly challenging, but there are numerous ways to simplify the process. One way to reconcile a large quantity of complex environmental data for processes and systems is via life cycle assessment. The life cycle assessment framework can help establish a quantitative description of the impact of a product or service [4] and it has been used to aid decision-making for a range of scenarios. Accordingly, life cycle assessment has been applied extensively to decisions pertaining to waste management e.g., [5,6,7,8]. In particular, there have been a number of studies that have addressed biosolids waste management e.g., [9,10,11].

However, no study to date has investigated biosolids reuse management in New Zealand where community dialogue has been included in the decision-making. Therefore, this study aims to calculate the environmental impact of the six identified biosolids reuse options using a life cycle assessment approach and a community dialogue mechanism. This information will be used in the final community hui to contribute to the decision making about the reuse options available.

8.2 MATERIALS AND METHODS

This study has been conducted in accordance with the principles and framework detailed in the ISO standards on life cycle assessment [12,13]. The goal of the study was to assess the potential environmental impacts for the biosolid reuse options preferred by the Kaikōura community. The functional unit is the treatment of one ton of the stockpiled Kaikōura biosolids. This assessment extends from the biosolids, as they are stockpiled now, to the end of life associated with the various reuse options. Accordingly, the system boundaries extend from the extraction of raw materials to their eventual disposal.

A community hui involving the district council, tangata whenua ("people of the land") and community group representatives was held in the Takahanga marae in Kaikōura in December 2011. Here we explained the environmental impacts quantified by this life cycle assessment study (see Table 1 for the impact categories used). Then each stakeholder was given ten votes numbered one to ten that they could allocate to each of the impact categories to represent how important the different environmental impacts were to the Kaikōura community regarding the biosolids reuse options. It was explained that the numerical value of the vote was proportional to the importance. The votes were subsequently used to generate weightings for each of the impact categories so they could be aggregated. Details of the environmental impacts and the reason for their inclusion are detailed in Table 1. In addition to the voting, stakeholders were encouraged to record the reasons for their vote and discuss their opinions about the different environmental impact and how or why they were or were not relevant to their community on this issue.

For the life cycle inventory, the landfill process that was used to draw relative impact assessments was from the Ecoinvent database [14]. The bulk density of the biosolids was measured and found to be approximately 500 kg/m3, and green waste was assumed to be 200 kg/m^3 and for both composting options they were combined in equal amounts. All truck transportation steps were assumed to have been using a 25–30 t payload Euro 3 Truck [15] that is 50% utilized. A number of machines are necessary for the reuse processes and their efficiency is dependent on the workload. It was calculated that the loader used 1.3 kg diesel/t biosolids processed, the shredder 0.0018 kg diesel/kg biosolids, the mixer 0.0014 kg diesel/

kg biosolids. The storage of worm juice was calculated as requiring 54 kWh/t/day [16,17]. The production of polyethylene (PE) film, polyvinyl chloride (PVC) and repacking process were from the Ecoinvent database [14]. The direct gaseous emissions from vermiculture and open-air composting were from experimental studies [18,19]. The chemical composition of leachate from open-air composting was from [20] and it assumed that 80% of the leachate was reused and all of the leachate from vermiculture compositing was reused. The process operations of the biochar mixture were adapted from Ibarrola et al., [21] and Hammond et al., [22] to suit New Zealand conditions. It was assumed that open-air composting took 60 days and the vermi-composting 21 days. In all land application options it was assumed that 0.5% of the mineral nitrogen was lost as dinitrogen monoxide [23] and the percentage chemicals were assumed to be lost via leaching when directly applied to land was calculated using a range of studies [1,24,25,26,27,28]. The emissions and resources used for the land filling process were from the Ecoinvent database [14]. The electricity mix for New Zealand was from the GaBi database [29] and the diesel mix for New Zealand was from prior work by Scion [30]. The infrastructure for all the reuse options was assumed to last for 21 years, in accordance with the assumptions made in the Ecoinvent database for similar facilities [14]. The chemical composition of the Kaikōura biosolids was from a range of chemical analysis studies of the biosolids [31,32]. The displaced fertilizer from the biosolids reuse options was calculated using a range of studies depending on the reuse options investigated [18,20,33,34,35,36].

The data was compiled and modeled using the software GaBi4.4 [29]. The weightings were applied to the environmental impact categories using Equation 1 to develop the final environmental impact score of the different reuse options.

$$Index_{RU} = \sum \left(\frac{RU_n * W_n}{L_n * W_n}\right) \quad (1)$$

Equation 1: The equation used to calculate the overall score for the reuse options. Where: RU = the normalized impact of the reuse option, L = the normalized impact of landfilling option, W = weighting factor, and n = environmental impact category.

Table 1. The impact categories investigated in this study, their description and the reason for their inclusion in this study.

Impact category	Description	Reason for inclusion
Global Warming Potential	The potential radiative forcing of greenhouse gas chemicals in a steady state atmosphere [37].	Global warming potential was included because climate change is a significant issue for all communities and Kaikōura has an EarthCheck status—which includes a carbon footprint assessment.
Acidification Potential	The propensity of a chemical to contribute H+ ions to a medium.	Acidification potential is included because the biosolids management is potentially a source of this sort of pollution.
Eutrophication Potential	The potential nutrification of watercourses [38].	Eutrophication potential is included because the biosolids are potentially a significant source of this sort of pollution—depending on how they are treated.
Ozone Layer Depletion Potential	The change in stratospheric ozone column in a steady state [38].	Ozone layer depletion was used because New Zealand already has a high Ultra-Violet index, it would be prudent to choose a technology that does not exacerbate this.
Volumetric water use	A simple accounting approach to the appropriation of water [39].	Water is essential to New Zealand economy and there are several cultural connotations of water use.

Table 1. Continued.

Land occupation	The occupation of land where the land is classified and characterized according to a Hemeroby coefficient [38].	Land use is important on the east coast of the South Island because of the cultural significance to Iwi and economic opportunities associated with land-based activities.
Freshwater Ecotoxicity Potential	The potential toxic effects of a chemical on freshwater ecosystems using the UNEP-SETEC USEtox model [40].	Freshwater toxicity is included because the biosolids may be a particular source of toxic compounds and the reuse option evaluations need to take this into consideration.
Marine Aquatic Ecotoxicity Potential	The toxic effect on saltwater ecosystems and species [38].	Marine ecotoxocity is included because whale watching and marine tourism are significant sources of revenue for the Kaikōura community.
Human Toxicity Potential	The potential toxic effects of a chemical on human wellbeing using the UNEP-SETEC USEtox model [40].	Human toxicity is essential to quantify because there is a permanent population in the Kaikōura region.
Terrestrial Ecotoxicity Potential	The toxic effect on land based ecosystems and species [38].	Terrestrial toxicity needs to be included because the biosolids and reuse options may be a source of toxic compounds.

The most up-to-date and complete impact assessment methods were chosen for each of the impact categories and these are specified in Table 1. Each impact category is characterized by a particular chemical or concept. Global Warming Potential is converted to carbon dioxide equivalents (kg CO_2-eq), Ozone Depletion Potential to the refrigerant R11 (kg R11-eq), Eutrophication Potential is measured in phosphate equivalents (kg PO_4^{3-}–eq), Acidification Potential to sulfur dioxide equivalents (kg SO_2-eq), Marine and Terrestrial Ecotoxicity Potential to dichlorobenzene (kg DCB-eq), Human Toxicity Potential is measured by the number of cases of illness, Freshwater Toxicity Potential by the Potential Affected Fraction per unit volume and time (PAF.m^3.day), volume of Water Use is calculated to weight (kg), and the Land Use is the area used per unit time (m^2.yr-eq).

Normalization was undertaken to remove the effect of aggregating values with different units. The normalization data was obtained from the Gabi4.4 database and corresponds to Organization for Economic Co-operation and Development data. Except for the water use which was calculated by dividing the daily water use of New Zealand [41] by the population [42]; the data used is detailed in Table 2.

8.3 RESULTS AND DISCUSSION

The ranking votes produced in the December 2011 hui produced some interesting results (Figure 1). A key output is that global warming was ranked the least important environmental impact and that water use, land use and water quality received the majority of the votes (Figure 1).

The stakeholders recorded despondency about environmental issues such as global warming and ozone depletion because these issues require a concerted and consistent international effort that has hitherto been absent. Consequently, they felt the most pertinent issues for the Kaikōura community were the ones they could do something constructive about. Therefore land use, water use and water quality metrics scored the most important (Figure 1). Given that the main industries in Kaikōura involve farming or marine tourism, this is understandable. However, climate and water are inextricably linked and perhaps this approach does not make that explicitly clear. Given the diversity of stakeholders that attended the hui in

Table 2. The normalization data used for the impact categories included in this study.

Impact Category	Normalization data	Units
Global Warming Potential	5.26×10^{-14}	kg CO_2-eq
Ozone Depletion Potential	9.60×10^{-9}	kg R11-eq
Eutrophication Potential	1.42×10^{-11}	kg $PO4^{3}$-eq
Acidification Potential	9.59×10^{-12}	kg SO_2-eq
Marine Toxicity Potential	8.12×10^{-15}	kg DCB-eq
Terrestrial Toxicity Potential	1.62×10^{-12}	kg DCB-eq
Human Toxicity Potential	0.28×10^{1}	cases
Freshwater Toxicity Potential	8.33×10^{-13}	PAF.m^3
Water Use	1.88×10^{1}	kg
Land Use	2.86×10^{-4}	m^2.yr-eq

December, it would be expected that the opinions on the importance of the environmental impact categories would be diverse. However, we found an unexpected degree of consistency in the recorded opinions of the different impact categories—as indicated by the standard error of the mean bars (Figure 1).

The use of weighting is contentious in life cycle assessment and it has stimulated a substantial number of journal papers e.g., [43,44,45,46,47]. Finnveden et al., [48] conclude that there is no definitive way to weight in life cycle assessment and all methods suffer at least two shortfalls. Firstly, you cannot determine if the weighting approach accurately reflects the decision makers' values at that particular point in time, and secondly, any weighting method may conceal crucial assumptions. During this study we found that there can be significant discrepancy in the general public's understanding of environmental problems and the complexity therein. A key future objective would be to evaluate a range of weighting procedures via a sensitivity analysis—but this is out of the scope of this particular study. Weighting is particularly suitable for biosolids management decision making because the impact of each management options extends beyond a single environmental consequence and the overall impacts are

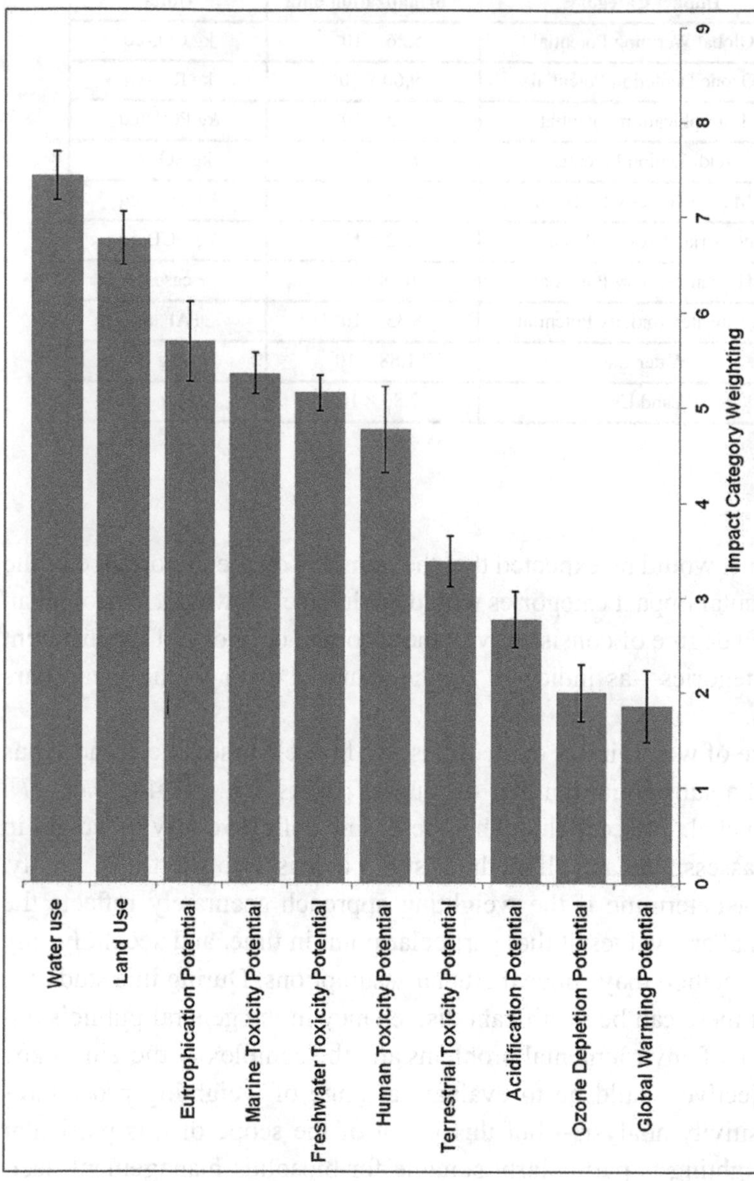

Figure 1. The environmental impact category weightings developed from the panel voting hui (error bars are the standard error of the mean). 10 = most important, 1 = least important.

disparate (Table 3). Resulting in a wealth of complicated and technical information, which viewed without aggregation may result in weighting by stealth. The integration of the Kaikōura community preferences regarding environmental impact into the analysis was a key feature of this work, and the feedback received during the hui was that this was an empowering process that gave the community more ownership and understanding of the analysis.

The data detailed in Table 3 corresponds to the environmental footprints calculated for the life cycle assessment metrics in this study. The footprints exhibit a disparate and complex trend between metrics and reuse options. Because the findings of other studies are heavily influenced by system boundaries, assumptions and underlying data, it is difficult to draw comparisons. However the trend exhibited between the options is intuitively understandable. For example, global warming potential shows a higher footprint for the composting options than the direct land application options. This is because of the gaseous emissions, materials and transportation necessary for the production of compost [18]. Similarly, land use is significantly more for the composting options than the direct land applications because of the infrastructure associated with the composting facilities. Water use and eutrophication is higher for the composting facilities because of the associated wastewater treatment necessary to treat the leachate. Conversely the freshwater and human toxicity metrics are higher for the direct to land application options because the biosolids are ostensibly unprocessed.

The data was combined to a single figure using Equation 1, and the overall scores reveal an interesting trend. As depicted in Figure 2, all the reuse options were found to have a lower environmental score than landfilling the biosolids. This is largely because the reuse options avoid the production of fertilizers and pollution and resource use associated with landfills. The options that involve the direct application to land are the most environmentally benign. However there is a relatively small difference in aggregated environmental impact between these options. What difference there is can be attributable to the use of heavy machinery and resources involved in incorporating the biosolids in forest soils; compared to a simple tractor-drawn spreader used for pastoral land application.

Table 3. Environmental impact scores for each of the biosolid reuse options and landfilling.

Impact Category	Unit	Landfill	Open Air Composting	Vermi-Com-posting	Mixture with Biochar	Pastoral Farmland	Exotic Forest Application	Land Reha-bilitation
Global Warming	kg CO_2-eq	7.00×10^{2}	3.06×10^{3}	2.03×10^{3}	3.15×10^{2}	2.85×10^{2}	8.64×10^{1}	8.65×10^{1}
Ozone Depletion	kg R11-eq	3.38×10^{-6}	7.26×10^{-6}	7.80×10^{-7}	5.40×10^{-7}	1.72×10^{-7}	2.03×10^{-7}	2.25×10^{-7}
Eutrophication	kg $PO4^{3-}$eq	0.76×10^{1}	2.26×10^{-1}	4.42×10^{2}	2.21×10^{2}	0.10×10^{1}	4.60×10^{-1}	4.60×10^{-1}
Acidification	kg SO_2-eq	2.98×10^{-1}	7.20×10^{-1}	3.09×10^{-1}	1.60×10^{-1}	0.32×10^{1}	6.60×10^{-2}	6.68×10^{-2}
Marine Ecotoxicity	kg DCB-eq	9.38×10^{5}	2.56×10^{4}	2.05×10^{4}	1.33×10^{4}	6.87×10^{2}	3.30×10^{2}	2.45×10^{2}
Terrestrial Ecotoxicity	kg DCB-eq	0.15×10^{1}	1.19×10^{-1}	1.24×10^{-2}	6.11×10^{-3}	3.56×10^{-1}	1.73×10^{-1}	1.74 × 10–1
Human Toxicity	cases	7.32×10^{-10}	1.99×10^{-9}	7.59×10^{-10}	3.92×10^{-11}	7.16×10^{-10}	8.95×10^{-12}	1.64×10^{-11}
Freshwater Toxicity	PAF.m^3.day	6.08×10^{-2}	2.13×10^{-1}	5.48×10^{-2}	3.57×10^{-2}	9.59×10^{-1}	1.45×10^{-2}	1.49×10^{-2}
Water Use	kg	31.3×10^{1}	0.48×10^{1}	0.22×10^{1}	0.15×10^{1}	5.31×10^{-1}	6.23×10^{-1}	6.23×10^{-1}
Land Use	m^2.yr-eq	2.60×10^{-4}	5.92×10^{-1}	4.23×10^{-1}	1.53×10^{-2}	6.38×10^{-6}	8.16×10^{-8}	1.70×10^{-7}

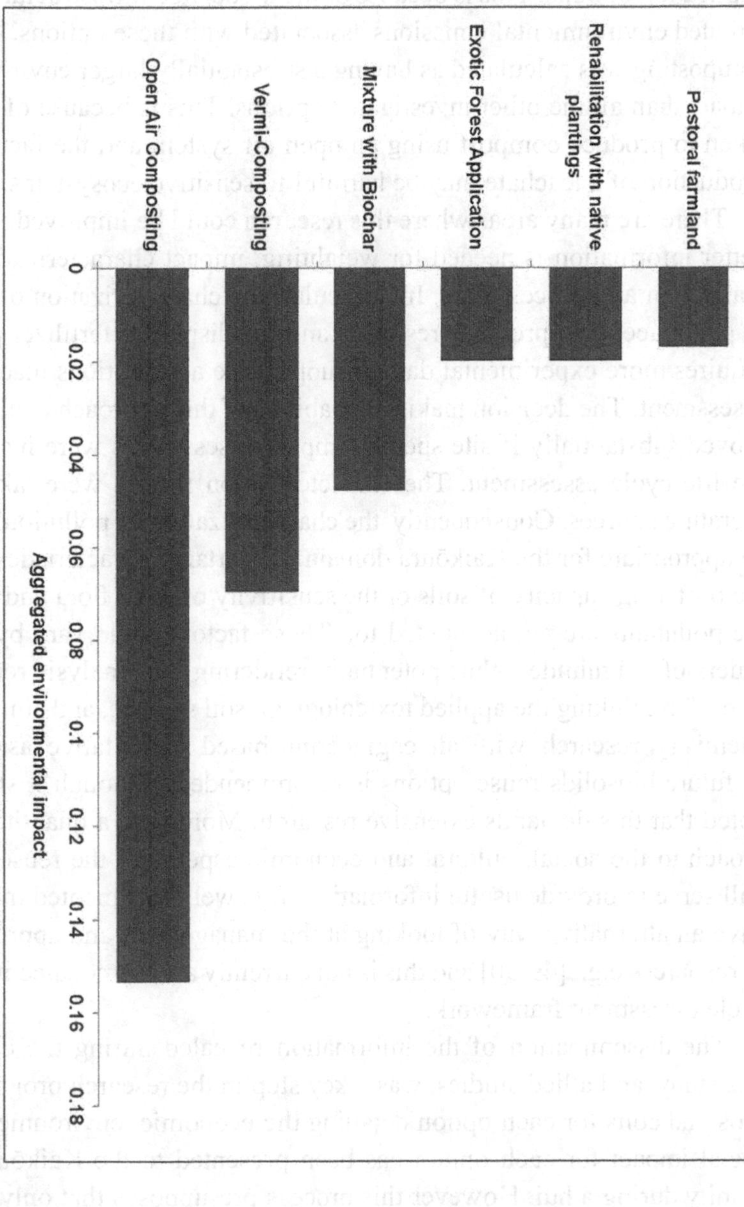

Figure 2. The aggregated environmental index for each of the reuse options presented relative to landfilling the waste, which would have an aggregated environmental impact score of 1.

The composting and the mixture with biochar options exhibit an increased aggregated environmental impact compared with the direct land application options. This is because of the associated infrastructure and facilitated environmental emissions associated with these options. Open air composting was calculated as having a substantially larger environmental impact than all the other investigated options. This is because of the time taken to produce compost using an open air system and the fact that the production of a leachate may be harmful to sensitive ecosystems.

There are many areas where this research could be improved. Notably, better information is needed for weighting, impact characterization, normalization and process data. In particular, the characterization of toxicity of pharmaceutical products residuals and the displaced fertilizer products requires more experimental data to support the assumptions made in this assessment. The decision making capability of this approach could be improved substantially if site specific impact assessments were included in the life cycle assessment. The characterization factors were taken from literature sources. Consequently, the characterization of pollution may not be appropriate for the Kaikōura domain. Important characteristics such as the buffering capacity of soils or the sensitivity of local flora and fauna to the pollutants are not accounted for. These factors could vary by several orders of magnitude—thus potentially rendering this analysis redundant. Therefore, linking the applied toxicological, soil science, and atmospheric chemistry research with an engineering-based quantitative assessment of future biosolids reuse options is recommended; although it should be noted that this demands extensive research. Moreover, a quantitative approach to the social, cultural and economic aspects of the reuse options will serve to provide useful information. It is well documented that Māori have an alternative way of looking at the management and appropriation of resources e.g., [49,50] and this is not currently accommodated in the life cycle assessment framework.

The dissemination of the information revealed during the course of this study, and allied studies, was a key step in the research program. The pros and cons for each option detailing the economic, environmental and social impact for each option has been presented to the Kaikōura community during a hui. However this process presupposes that only one option can be adopted and the Kaikōura community is obliged to manage their biosolid waste in isolation of other communities. Rural communities across the South Island of New Zealand are faced with similar challenges and a collaborative approach to biosolids management may reveal new opportunities and options. The options that involve significant infrastructure may become more environmentally benign due to a potential increase in throughput i.e. due to an economy of scale.

8.4 CONCLUSION AND FUTURE RECOMMENDATIONS

Making decisions that concern complex systems such as the natural environment is difficult and a quantitative approach can clarify the issues, especially for non-technical stakeholders. The results of this study suggest that the direct land application options have a relatively benign environmental impact compared to the options that involve significant infrastructure or reprocessing. This is likely to be symptomatic of the chemical composition of Kaikōura's biosolids and the total amount of biosolids to be processed considering the infrastructure necessary for the reuse options that involve reprocessing. Consequently, the findings of this study may not transpose to other regions, communities or times.

The community engagement aspect was a particular success and we advocate this approach with other communities on similar issues. Notably there was an acceptance of the life cycle assessment methodology by the Kaikōura community despite knowledge of its limitations. Moreover, the decision making capacity associated with life cycle assessment will be improved if the aforementioned limitations are addressed.

There has been reluctance from district councils in New Zealand to engage with communities on waste management decision making, but we have found that community engagement is a positive process that may reduce the risk associated with local council waste management decisions. However a key risk associated with this approach is that the overall assessment of options presupposes that a single reuse option will be favored, and that Kaikōura has to deal with their biosolids on their own. Perhaps a combination of options or a collaborative approach to biosolids management may serve the Kaikōura community best, and thus this warrants significant further investigation.

REFERENCES

1. Fitzmorris, K.B.; Sarmiento, F.; O'Callaghan, P. Biosolids and sludge management. Water Environ. Res. 2009, 81, 1376–1393.
2. Magesan, G.N.; Wang, H.; Clinton, P.W. Best Management Practices for Applying Biosolids to Forest Plantations in New Zealand; Scion Internal Report 17514; Scion: Christchurch, New Zealand, 2010.
3. Yusuf, S.A.; Georgakis, P.; Nwagboso, C. Procedural lot generation for evolutionary urban layout optimization in urban regeneration decision support. ITCON 2011, 16, 357–380.
4. Baumann, H.; Tillman, A.-M. The Hitchhikers guide to LCA. An Orientation in Life Cycle Assessment Methodology and Application, 1st ed.; Studentlitteratur AB: Lund, Sweden, 2004; p. 543.
5. Gunamantha, M. Sarto Life cycle assessment of municipal solid waste treatment to energy options: Case study of KARTAMANTUL region. Yogyakarta. Renew. Energ. 2012, 41, 277–284.
6. Koroneos, C.J.; Nanaki, E.A. Integrated solid waste management and energy production-A life cycle assessment approach: The case study of the city of Thessaloniki. J. Clean. Prod. 2012, 27, 141–150.
7. Curry, R.; Powell, J.; Gribble, N.; Waite, S. A streamlined life-cycle assessment and decision tool for used tyres recycling. Proc. Inst. Civ. Eng. 2011, 164, 227–237.
8. Tunesi, S. LCA of local strategies for energy recovery from waste in England, applied to a large municipal flow. Waste Manage. 2011, 31, 561–571.
9. Peters, G.M.; Rowley, H.V. Environmental comparison of biosolids management systems using life cycle assessment. Envir. Sci. Tech. Lib. 2009, 43, 2674–2679.
10. Sablayrolles, C.; Gabrielle, B.; Montrejaud-Vignoles, M. Life Cycle Assessment of Biosolids Land Application and Evaluation of the Factors Impacting Human Toxicity through Plant Uptake. J. Ind. Ecol. 2010, 14, 231–241.
11. Peters, G.M.; Lundie, S. Life-cycle assessment of biosolids processing options. J. Ind. Ecol. 2001, 5, 103–121.
12. ISO14040: 2006. Environmental Management-Life Cycle Assessment-Goal and scope definition and inventory analysis, International Organization for Standardization, Geneva, Switzerland, 2006.
13. ISO14044:2006. Environmental Management-Life Cycle Assessment-Requirements and Guidelines, International Organization for Standardization, Geneva, Switzerland, 2006.
14. Frischknecht, R.; Jungbluth, N.; Althaus, H.J.; Doka, G.; Dones, R.; Heck, T.; Hellweg, S.; Hischier, R.; Nemecek, T.; Rebitzer, G.; Spielmann, M. The ecoinvent database: Overview and methodological framework. Int. J. Life Cycle Ass. 2005, 10, 3–9.
15. MoT. The New Zealand vehicle fleet: Annual fleet statistics, 2009. In Ministry of Transport, Te Manatu Waka; A statistical report; Wellington, New Zealand, 2010. ISBN 978-0-478-07228-0.
16. Thompson, J.; Singh, P. Status of Energy Use and Conservation Technologies Used in Fruit and Vegetable Cooling Operations in California; California Energy

Commission, PIER Program, CEC-400-1999-005; University of California: Davis, United States of America, 2008.
17. McLaren, S., Jr.; Love, R.; McDevitt, J.E. Life Cycle Assessment Data Sets Greenhouse Gas Footprinting Project inventory report: Coolstores. A report for MAF and Zespri International (No. 12247); Ministry for Agriculture and Forestry: Wellington, New Zealand, 2011.
18. Chan, Y.C.; Sinha, R.K.; Wang, W. Emission of greenhouse gases from home aerobic composting, anaerobic digestion and vermicomposting of household wastes in Brisbane (Australia). Waste Manage. Res. 2011, 29, 540–548.
19. Rodriguez, V.; Valdez-Perez, M.D.L.A.; Luna-Guido, M.; Ceballos-Ramirez, J.M.; Franco-Hernandez, O.; van Cleemput, O.; Marsch, R.; Thalasso, F.; Dendooven, L. Emission of nitrous oxide and carbon dioxide and dynamics of mineral N in wastewater sludge, vermicompost or inorganic fertilizer amended soil at different water contents: A laboratory study. Appl. Soil Ecol. 2011, 49, 263–267.
20. Forgie, D.J.L.; Sasser, L.W.; Neger, M.K. Compost Facility Requirements Guideline: How to Comply with Part 5 of the Organic Matter Recycling Regulation; Ministry of Water Land and Air Protection: British Columbia, Canada, 2004.
21. Ibarrola, R.; Shackley, S.; Hammond, J. Pyrolysis biochar systems for recovering biodegradable materials: A life cycle carbon assessment. Waste Manage. 2012, 32, 859–868.
22. Hammond, J.; Shackley, S.; Sohi, S.; Brownsort, P. Prospective life cycle carbon abatement for pyrolysis biochar systems in the UK. Energ. Policy 2011, 39, 2646–2655.
23. IPCC. IPCC Guidelines for National Greenhouse Gas Inventories: Volume 4: Agriculture, Forestry and other Land Use; Intergovernmental Panel on Climate Change: Paris, France, 2006. Available online: http://www.ipcc-nggip.iges.or.jp/public/2006gl/vol4.htm (accessed on 21 August 2012).
24. Adam, K. The Environmental and Health Implications of the Decomposition of Biosolids; University of Canterbury: Christchurch, New Zealand, 2003.
25. Palma, R.M. Evaluation of Ammonia volatilisation, Carbon Dioxide evolution and N balance from biosolids following application to forest soils. M.S. Thesis, University of Canterbury, Christchurch, New Zealand, 2000.
26. Knowles, O.A.; Robinson, B.H.; Contangelo, A.; Clucas, L. Biochar for the mitigation of nitrate leaching from soil amended with biosolids. Sci. Total Environ. 2011, 409, 3206–3210.
27. Brown, S.; Beecher, N.; Carpenter, A. Calculator tool for determining greenhouse gas emissions for biosolids processing and end use. Envir. Sci. Tech. Lib. 2010, 44, 9509–9515.
28. Pierzynski, G.M.; Gehl, K.A. Plant nutrient issues for sustainable land application. J. Environ. Qual. 2005, 34, 18–28.
29. PE International. GaBi 4.4 Professional Life Cycle Software; University of Stuttgart: Germany, 2009. Available online: http://www.gabi-software.com (accessed on 3 May 2010).
30. McDevitt, J.E.; Seadon, J. Life Cycle Assessment Data Sets Greenhouse Gas Footprinting Project: Diesel. A report prepared for MAF and Zespri International (No. 12247); Ministry for Agriculture and Forestry: Wellington, New Zealand, 2011.

31. Robinson, B. Chemical Composition of the Kaikoura Biosolids. Hui presentation at Takahanga marae. April 2011. Unpublished work.
32. Northcott, G. Contaminants in the Kaikoura Biosolids. Hui presentation at Takahanga marae. April 2011. Unpublished work. [Google Scholar]
33. Cadena, E.; Coln, J.; Artola, A.; Sanchez, A.; Font, X. Environmental impact of two aerobic composting technologies using life cycle assessment. Int. J. Life Cycle Ass. 2009, 14, 401–410.
34. van Haaren, R.; Themelis, N.J.; Barlaz, M. LCA comparison of windrow composting of yard wastes with use as alternative daily cover (ADC). Waste Manage. 2010, 30, 2649–2656.
35. Fernandez-Luqueao, F.; Reyes-Varela, V.; Martanez-Suarez, C.; Reynoso-Keller, R.E.; Mandez-Bautista, J.; Ruiz-Romero, E.; Lapez-Valdez, F.; Luna-Guido, M.L.; Dendooven, L. Emission of CO2 and N2O from soil cultivated with common bean (Phaseolus vulgaris L.) fertilized with different N sources. Sci. Total Environ. 2009, 407, 4289–4296.
36. IPCC. Solid Waste disposal. 2006 IPCC Guidelines for National Greenhouse Gas Inventories. Prepared by the National Greenhouse Gas Inventories Programme; Hayama, Japan, 2006.
37. IPCC. Climate Change 2007. IPCC Fourth Assessment Report. The Physical Science Basis; 2007.
38. Guinée, J.B. Handbook on life cycle assessment. Operational guide to ISO standards; Kluwer Academic Publishers: Dordrecht, The Netherlands, 2002; p. 692.
39. Hoekstra, A.Y.; Chapagain, A.K.; Aldaya, M.; Mekonnen, M.M. Water Footprint Manual-State of the Art; Water Footprint Network: Enschede, The Netherlands, 2009.
40. Rosenbaum, R.K.; Bachmann, T.M.; Gold, L.S.; Huijbregts, M.A.J.; Jolliet, O.; Juraske, R.; Koehler, A.; Larsen, H.F.; MacLeod, M.; Margni, M.; et al. USEtox-The UNEP-SETAC toxicity model: Recommended characterisation factors for human toxicity and freshwater ecotoxicity in life cycle impact assessment. Int. J. Life Cycle Ass. 2008, 13, 532–546.
41. StatsNZ. Water Physical Stock Account: 1995-2005; Statistics New Zealand: Wellington, New Zealand, 2007.
42. StatsNZ. Census of Population and Dwellings - Final Counts; Statistics New Zealand: Wellington, New Zealand, 2006.
43. Reap, J.; Roman, F.; Duncan, S.; Bras, B. A survey of unresolved problems in life cycle assessment. Part 2: Impact assessment and interpretation. Int. J. Life Cycle Ass. 2008, 13, 374–388.
44. Johnsen, F.M.; Løkke, S. Review of criteria for evaluating LCA weighting methods. Int. J. Life Cycle Ass. 2012, 1–10.
45. Yellishetty, M.; Ranjith, P.G.; Tharumarajah, A.; Bhosale, S. Life cycle assessment in the minerals and metals sector: A critical review of selected issues and challenges. Int. J. Life Cycle Ass. 2009, 14, 257–267.
46. Finnveden, G.; Eldh, P.; Johansson, J. Weighting in LCA based on ecotaxes: Development of a mid-point method and experiences from case studies. Int. J. Life Cycle Ass. 2006, 11, 81–88.

47. Koffler, C.; Schebek, L.; Krinke, S. Applying voting rules to panel-based decision making in LCA. Int. J. Life Cycle Ass. 2008, 13, 456–467.
48. Finnveden, G.; Hofstetter, P.; Bare, J.; Basson, L.; Ciroth, A.; Mettier, T.; Seppälä, J.; Johansson, J.; Norris, G. Normalisation, grouping, and weighting in life cycle impact assessment. In Life Cycle Impact Assessment: Striving Towards Best Practice. Society of Environmental Toxicology and Chemistry (SETAC); de Haes, H.A.U., Ed.; Pensacola, FL, USA, 2002.
49. Tipa, G.; Teirney, L. Cultural Health Index for Streams and Waterways: A tool for nationwide use. A report prepared for the Ministry for the Environment (No. 710); Ministry for the Environment: Wellington, New Zealand, 2006.
50. Rotarangi, S.; Thorp, G. Can profitable forest management incorporate community values? New Zeal. J. For. 2009, 54, 13–16.

PART IV

POLICY PLANNING FOR THE FUTURE

CHAPTER 9

Policy Instruments Towards a Sustainable Waste Management

GÖRAN FINNVEDEN, TOMAS EKVALL,
YEVGENIYA ARUSHANYAN, MATTIAS BISAILLON,
GREGER HENRIKSSON, ULRIKA GUNNARSSON ÖSTLING,
MARIA LJUNGGREN SÖDERMAN, JENNY SAHLIN,
ÅSA STENMARCK, JOHAN SUNDBERG, JAN-OLOV SUNDQVIST,
ÅSA SVENFELT, PATRIK SÖDERHOLM, ANNA BJÖRKLUND,
OLA ERIKSSON, TOMAS FORSFÄLT, AND MONA GUATH

9.1 INTRODUCTION

The global community is facing several environmental challenges (e.g., [1,2]). Climate change, loss of biodiversity, disrupted biogeochemical cycles and use of hazardous substances are examples of environmental problems threatening a sustainable development. Fourteen out of the sixteen Swedish Environmental Quality Objectives, defining the environmental dimension of sustainable development, will not be met unless new policy measures are taken [3]. In order to develop in a more sustainable direction, all sectors of society, including waste management, need to implement measures that can lead towards a more sustainable society. The generation and management of waste depends on what activities are going on in society, and also on how these activities are controlled by public authority. In order to control the activities, decision-making bodies implement specific policy instruments, as well as issue documents, stating general policy objectives.

Responding to both economic and environmental challenges, the European Commission [1] has developed a road map for a resource efficient Europe. For waste management, the road map sets out several milestones for 2020, including:

- Waste generated per capita is in absolute decline.
- Energy recovery is limited to non-recyclable materials.
- Landfilling is virtually eliminated.
- High quality material recycling is ensured.

The waste management sector has a unique possibility of not only reducing its own environmental impacts, but it can also, through increased utilization of waste, contribute to other sectors' emission reductions. It has also been shown that an environmentally optimized waste management system can have significantly lower overall environmental impacts than the current system (e.g., [4,5,6]). Treatment of solid waste is surrounded by a number of rules, regulations and policy instruments. These may be quite different in different European countries [7,8] depending on traditions and contexts. The environmental impacts from the waste management systems are also quite different in different countries [9].

Swedish waste policy depend on a number of policy documents, including the European Union waste directive, Swedish environmental quality objectives, and policies in other sectors, including the energy sector. The European waste directive requires that the waste hierarchy should be used although exemptions can be made based on life-cycle thinking [10]. The waste hierarchy states that waste should be managed in a priority order, from prevention; to preparing for re-use; to recycling; to other recovery (e.g., energy recovery) and to the final option disposal. The Swedish environmental objective for achieving a "good built environment" states that waste disposal should be efficient for society and convenient for consumers and that waste is prevented, resources in the waste are used as much as possible while the impacts and risks for the environment and human health are minimized [11]. Waste management is also important for achieving several other environmental quality objectives including "reduced climate impact" and "a non-toxic environment" (ibid.).

Waste management in Sweden and in many other countries has undergone significant changes during the last decades. Figure 1 describes the development for household wastes indicating the clear increase in incineration and recycling and a resulting decrease in landfilling.

In 2010 a total of 117.6 million tons of waste were generated in Sweden. 2.5 million tons were classified as hazardous waste [14]. 4.2 million tons of total waste was the so-called secondary waste generated by waste treatment. The industrial sector of mining and quarrying (mining) accounted for 89 million tons of waste, and waste from other manufacturing industry for 7.8 million tons. The construction sector generated 9.4 million tons of waste while the infrastructure sector (energy and water supply, and sewerage and sanitation) generated 1.7 million tons. Households generated more than 4 million tons, Services generated 1.8 million tons and Agricultural industries (forestry, agricultural and fishing industries) around 310,000 tons of waste. Waste treatment generated 3.5 million tonnes of waste.

About 80% of all waste was landfilled [14]. If mining waste is excluded, 43% of remaining waste was recycled, 28% was used as fuel, 13% was landfilled, and 16% was disposed by land treatment or discharged to water. Recycling includes conventional material recycling (for example of paper, metals, glass and plastics), biological treatment and the use of construction materials and materials for landfill cover.

Swedish policy instruments affecting the waste management system [15] include a ban on landfill disposal of organic materials, a landfill tax and an extended producer responsibility of some product groups, including packaging waste and wastes of electrical and electronic equipment. In addition, there are also energy and carbon dioxide taxes on fossil fuels used for heating. These policy instruments have overall been effective in influencing behavior and waste management has changed.

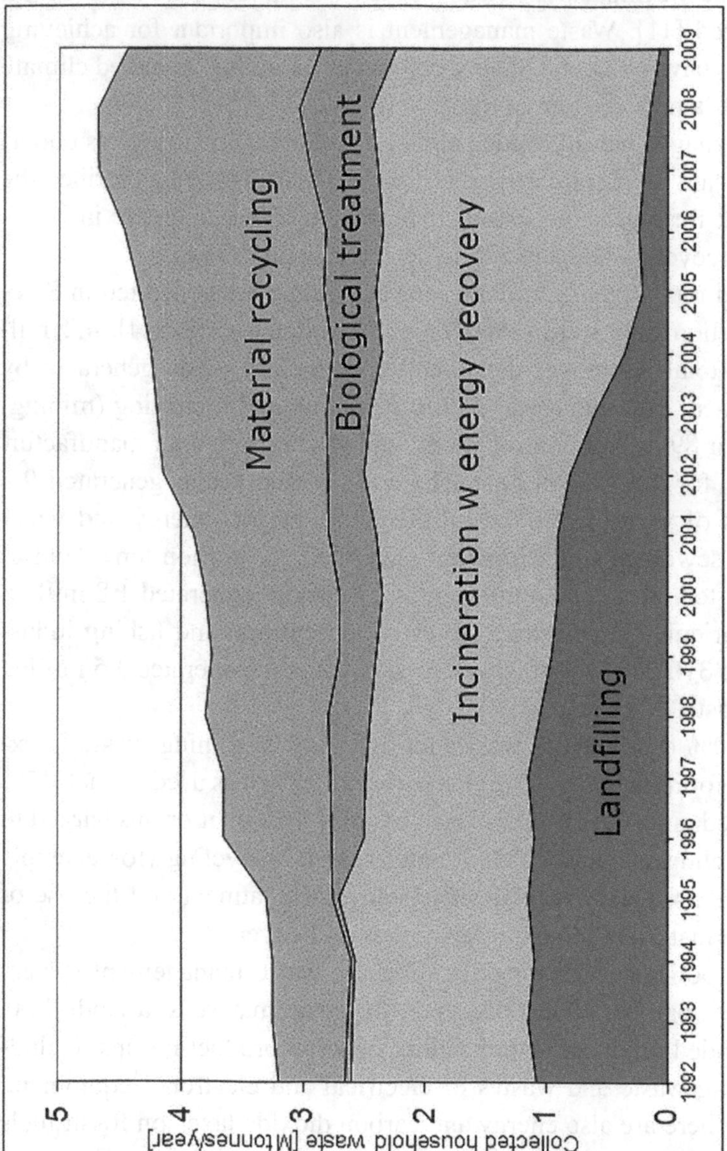

Figure 1. Treatment of collected municipal solid waste household waste in Sweden [12,13].

It can be noted that most legislation operating in the field is moving waste away from landfill disposal. There are currently only a few general policy instruments that support waste prevention and increased re-use and recycling, in order to promote the higher levels of the waste hierarchy. One example is the extended producer responsibility, but it includes only a limited number of waste fractions and it does not require any recycling above the target level. To comply with the waste hierarchy there is thus a need for new policy instruments. It can also be noted that waste prevention aims not only at reducing the amounts of waste, but also at reducing the hazardousness of the waste and the environmental impacts from treatment of the waste, which suggests that policy instruments, focusing on waste prevention, should not only address waste reduction. This implies, for instance, that policy instruments in the chemicals field may have important positive impacts in this regard. Furthermore, as individual choices and socially constructed and maintained habits determine the potential for achieving sustainable waste management, policy measures promoting individuals, in households as well as in workplaces, to recycle are also needed [16,17].

The waste management system is strongly integrated in other parts of society. Thus, policies and policy instruments in other sectors will also influence the waste management. For example, waste incineration accounts for 16% of the district heating produced in Sweden [18]. All policies and policy instruments within the energy sector will therefore indirectly also influence the waste management sector. Since the energy sector is influenced by a number of policies affecting, for example, climate change, energy security and industrial competitiveness, new and existing policy instruments for the energy sector are likely to evolve.

In order to develop more sustainable waste management systems, policy instruments are needed, not the least instruments that can support the higher levels of the waste hierarchy and address the complexity of the waste management system. With the purpose to fill these policy gaps, and suggest new policy instruments for a more sustainable waste management the multi-disciplinary Swedish research program "Towards a sustainable waste management" (TOSUWAMA) was initiated by the

Swedish EPA. One of the aims of the research program has been to identify and evaluate new policy instruments. The program involves nine Swedish research partners (see [19]). In the program, a more sustainable waste management system is defined as a system that contributes to increasing efficiency in the use of natural resources, and to decreasing environmental burdens. Furthermore, environmental improvements within Sweden should not be offset by unwanted consequences in other countries. To be sustainable, the waste management must also be affordable and widely accepted by the public as well as by key companies and organizations. In the program, the policy instruments intended for sustainable waste management have been evaluated in several parallel studies looking into economic aspects, environmental impacts and the social acceptance of the policy instruments.

The aim of this paper is to suggest and discuss policy instruments that could lead us towards a more sustainable waste management, along with proposals for further development. The paper is heavily based on evaluations from the research program, but in part it also draws from the results of other research studies. The paper is thus a synthesis of more detailed studies where specific policy instruments have been analyzed using specific methods. By making this broad synthesis we are able to draw conclusions that are not possible when more detailed studies are presented. Given the strong multi-disciplinary focus the paper does not provide a full-fledged overview of the existing literature, and/or detailed methodological descriptions. The presentation is brief emphasizing results, and the reader will need to consult the separate studies in the program for further details. The primary target group for our research is the Swedish government and authorities. For this reason, the primary focus of the assessment is on policy instruments that the Swedish government and authorities can decide on, in other words on a separate, Swedish implementation of the instruments. However, a broader geographical scope is also relevant, since Swedish authorities can choose to strive for the implementation of some of the policy instruments on, for example, the EU level. Although not our primary target group, most of the content in this paper should also be relevant for policy makers in other industrialized countries around the world.

9.2 METHODS

9.2.1 INTRODUCTION

This paper synthesizes and draws conclusions from several empirical studies made within the TOSUWAMA research program. These assessments are published in other reports and papers, which are used as references in this publication. Besides results from the program, also other relevant results are referred to in the discussion.

Bisaillon et al. [20] and Finnveden et al. [21] presented an inventory of a large number of policy instruments suggested by stakeholders and in the literature. Based on this inventory they identified 16 instruments as interesting candidates deserving further evaluation. This identification was based on the results from a workshop with stakeholders but also criteria developed within the program. The criteria for choosing the interesting candidates included environmental and economic impacts and social acceptability, but also program-specific criteria such as novelty and research interest. In the research program 13 of the 16 policy instruments (Table 1) have been assessed from three main perspectives: economic impacts, environmental impacts and social acceptance. In addition, a futures perspective was taken. Specifically, each type of assessment was made with reference to different possible future developments, illustrated in five external scenarios for the year 2030 [22,23]. These scenarios are:

- 0: Reference scenario, assuming developments in accordance with official forecasts made in 2008
- 1: Global sustainability, assuming globalization and strong political control over the environment and natural resources.
- 2: Global markets, assuming globalization and weak political control over the environment and natural resources.
- 3: Regional markets, assuming regionalization and weak political control over the environment and natural resources.
- 4: European sustainability, assuming regionalization and strong political control over the environment and natural resources.

Table 1. Assessment of policy instruments in the research program TOSUWAMA.

Policy instrument	Economic assessment	Environmental assessment	Assessment of social acceptance
Climate tax on waste incineration	X	X	
Including waste in the green certificate system for electricity production	X	X	
Compulsory recycling of recyclable materials	(X)	(X)	
Tradable Recycling Credits	X		
Weight-based tax on incineration of waste	(X)	(X)	
Weight-based waste collection fee	X	X	X
Developed recycling systems			X
Tax on virgin raw materials	X	X	
Advertisements on request only		X	X
Differentiated VAT	X	X	
Environmentally differentiated waste fee			X
Information to household and enterprises			X
Mandatory labeling of goods containing hazardous substances			X

(X) indicates that the evaluations are based on previous studies.

Results from the evaluations of the policy instruments in Table 1 are presented in Section 3. In the discussion in Section 4 also other policy instruments (e.g., other instruments identified by [20]) are included.

9.2.2 INTEGRATED APPROACH FOR QUANTITATIVE ANALYSIS

Several methods and scientific disciplines have been applied in the assessment of policy instruments within the research program TOSUWAMA. For the quantitative analysis, three existing quantitative tools have been combined and refined in order to assess economic and environmental aspects [24,25]:

- The Environmental Medium term Economic model (EMEC) is a computable general equilibrium (CGE) model of the Swedish economy [26]. The EMEC model has been extended in order to analyze the relation between economic activity and waste generation. Data on waste quantities has been compiled and assigned to different economic activities and different sectors [23]. In the model, the waste generation of households and firms depend on their respective economic activities and is sensitive to changes in the price of goods and services. The waste-management costs are assumed to affect the total cost of utilizing goods and services. Hence, households and firms incorporate waste-management performance into their decisions [24]. The waste generation is directly or indirectly influenced by changes in government policies, e.g., tax policies [27].
- NatWaste is a systems engineering model of the Swedish waste management system [28,29]. Based on cost optimization, NatWaste calculates the cost-effective mix of technologies for managing Swedish waste. The cost-effective mix is the set of technologies that gives the lowest total economic costs (excluding external environmental costs and private consumers' time) on the basis of the conditions defined for the analysis. Among the most influencing conditions are the choice of treatment technologies defined for each waste type (including their unit costs and performance) as well as the scenarios.

- Swedish Waste management Environmental Assessment (SWEA) is a life cycle assessment (LCA) model of the Swedish waste-management system [30]. LCA is a tool for assessing the potential environmental impacts of a product or a service (e.g., [31]), in this case waste management. Since a life-cycle perspective is used, credit is given to useful products, materials and energy carriers produced in the waste-management system that can replace products produced from virgin raw materials, in line with established LCA methodology for waste management (e.g., [32,33]). In addition, SWEA includes the reductions in material production of material that follows from waste-prevention efforts. This allows the model to account for the environmental benefits of waste prevention. SWEA has been implemented in the Simapro software [34] and for Life Cycle Impact Assessment the Recipe methodology [35] was used together with Cumulative energy demand [36] and Cumulative exergy demand [37].

The three models feed each other with information (Figure 2). EMEC and NatWaste are soft-linked in the sense that some variables solved for in one model are transferred into the data set of the other model in an iterative process. The last step is to feed the cost-effective mix of waste management technologies as calculated by NatWaste into SWEA for analyzing the life cycle environmental impacts. The linking of these three models allows us to consider how policy instruments intended to prevent waste generation or direct waste management in a more sustainable direction could affect: (1) the macroeconomic development, such as GDP growth and structural changes in the economy as a whole, (2) the cost-effective mix of technologies for managing Swedish waste and (3) the resulting life cycle environmental impacts. Furthermore, the approach makes it possible to capture if and how waste-management costs affect waste generation [25].

One advantage of this integrated modeling approach is that it enables a broad analysis. A general equilibrium model covers the whole economy in a geographical area, and can thus address important interactions between different sectors in the economy in a consistent manner. NatWaste calculates net costs for managing many of

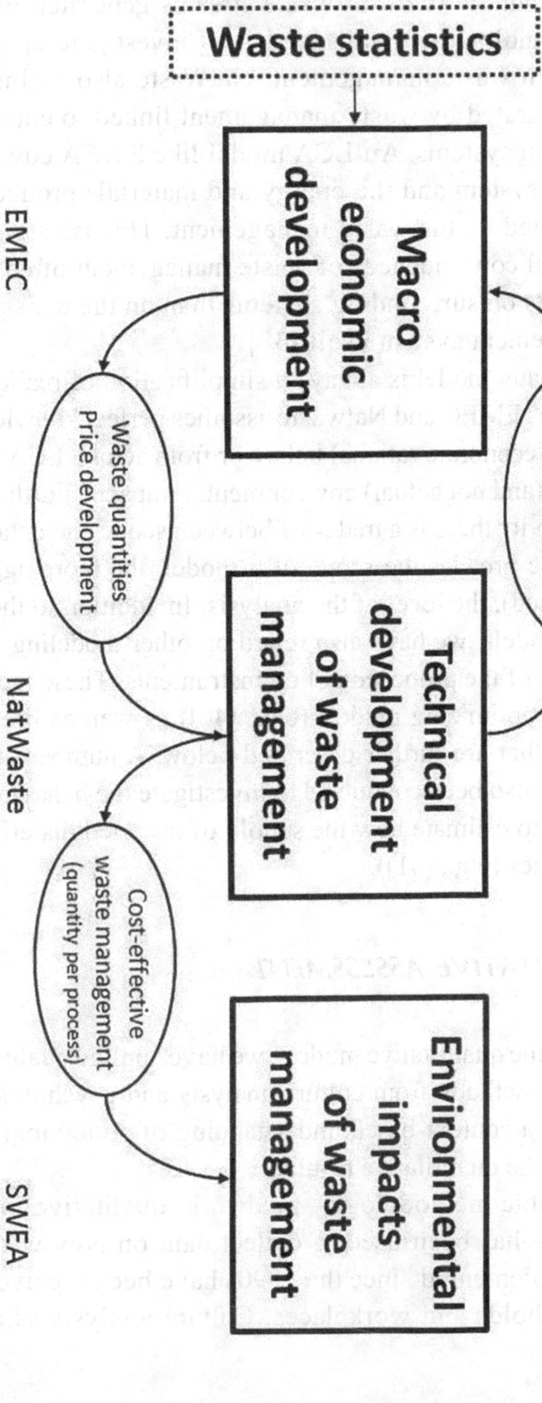

Figure 2. Combination of models for assessing policy instruments for sustainable waste management [25].

the environmentally relevant waste streams generated in Sweden and adds the technological detail needed to investigate specific technology choices in waste management. NatWaste also includes costs and revenues generated by waste management linked to energy and material production systems. An LCA model like SWEA covers the waste-management system and the energy and materials production systems that are affected by the waste management. This is essential since the environmental consequences of waste management often depend more on the impacts on surrounding systems than on the emissions from the waste management system itself [38].

However, any model is always a simplification of reality. Optimizing models such as EMEC and Natwaste assumes perfect knowledge about future costs and economic rational behavior from actors. LCA models calculate potential (and not actual) environmental impacts. Furthermore, in any modeling activity there is a trade-off between scope and detail [39], which means that the broader the scope of a model, the more aggregated (and thus generalized), the level of the analysis. In addition to the results from these three models, we have also relied on other modeling approaches in the evaluation of the proposed policy instruments. These include different economically optimizing models (e.g., [40]) as well as life cycle assessment models that are further described below. A number of econometric analyses have also been conducted to investigate the behavior of recycling markets, e.g., to estimate how the supply of recycled materials is affected by price changes (e.g., [41]).

9.2.3 QUALITATIVE ASSESSMENT

In addition to the quantitative models, we have applied qualitative analyses [42] and also methods from culture analysis and psychology [43,44,45]. This provides a context-based understanding of policy impacts, and thus complements the quantitative results.

Ethnographic methodology—fieldwork, qualitative interviews and observations—has been used to collect data on how waste policy instruments implemented since the 1990s have been received in the context of households and workplaces. Culture analysis as a method for analyzing ethnographic data (see, e.g., [46]) has been applied in order to find out how existing policy instruments, i.e., current waste handling conditions in everyday life, are anticipated, accepted, and acted upon. Issues analyzed include in what ways policy measures are based upon general cultural understandings concerning, for example, protection of the environment and also how different actors (e.g., municipalities) are seen as (economically, politically and morally) responsible for taking care of waste. Furthermore, we have analyzed whether waste legislation, as it is operating in everyday life contexts, has been perceived legitimate, and comprehensible (i.e., possible to understand in a meaningful way).

The qualitative evaluation of a handful of suggested (not yet implemented) waste policy instruments has been based on a combined approach of ethnology and psychology. The method for this qualitative ex-ante evaluation of new policy instruments was developed as part of the program [45]. It included several group discussions within the project team, one of them also including external laymen and stakeholders. The resulting evaluation was thus based on reasoning concerning whether each of the instruments were in line with cultural patterns and also psychological parameters. In practice the evaluation entailed the following analytical categories:

- (i) if the instrument matched the individual's / household's environmental commitment;
- (ii) the perceived social fairness of the instrument;
- (iii) if the policy instrument would affect the individuals or households directly or indirectly (i.e., through other stakeholders, such as landlords);
- (iv) how the policy instruments would interact, or be in conflict, with fundamental cultural categories and practices [43];
- (v) if the instrument would conform or not with the users' general understanding [47] of the waste system's task and function (e.g., to be a community service that minimizes environmental impact), and
- (vi) the (un)certainty of the message conveyed through the policy instrument (uncertainty regarding environmental impact; the

benefit of oneself doing something and uncertainty about what others are doing, i.e., social uncertainty).

Based on the methodology outlined, Andersson et al. [44] produced a summary for each instrument assessed, concerning how well it would function from a combined ethnological and psychological perspective. This study presented conclusions as to whether the policy instrument was considered socially and culturally anticipated and acceptable, and gave recommendations on how it could be modified to better achieve its purpose.

The qualitative research in TOSUWAMA also involves a stakeholder analysis [48] and a comparative static analysis of the impacts of developed recycling systems, tradable recycling credits and virgin materials taxes [49]. Such qualitative research is important as it explicitly addresses the key characteristics of different policy instruments (e.g., the incentives provided by a tax), and not only the impacts of specific policy proposals (e.g., a tax of x SEK/kg). This is also complemented by drawing on the practical experiences of policy instruments in the past and in other countries (e.g., [50]).

In sum, the emphasis on multi-disciplinary research efforts and the combination of quantitative and qualitative approaches, imply that a holistic approach to policy instrument evaluation is employed. In the following the most important insights and results from the program are outlined.

9.3 EVALUATION OF POLICY INSTRUMENTS

In the following, each of the evaluated policy instruments is described in terms of design and assumptions about its (hypothetical) implementation, along with results from the qualitative and quantitative assessments. Table 1 gives an overview of what type of assessments were done for each policy instrument as part of the projects within this research program. These assessments are complemented with findings from a number of related research projects (although no complete overview of the existing literature is provided).

9.3.1 CLIMATE TAX ON WASTE INCINERATION

9.3.1.1 DESCRIPTION AND ASSUMPTIONS

This policy instrument is a tax on fossil CO_2 emissions generated from waste incineration. In the analyses, the level of the tax is assumed to be 0.95 SEK/kg CO_2 for waste incineration plants with district heat production only (DHO) and 0.15 SEK/kg CO_2 for waste incineration plants with combined heat and power production (CHP) (1 SEK corresponds to about 0,11 EUR and 0.15 USD, respectively in February 2013). The differentiation of the tax level is a result of the Swedish energy taxation system with the aim of increasing combined heat and power production and it is further described by Bisaillon et al. [20]. The proposed policy instrument would imply that CO_2 from fossil sources in the waste is taxed in the same way as fossil fuels in general.

Most of the fossil CO_2 emissions originate from plastic waste. The evaluated tax levels correspond to 2900 SEK/ton plastic waste at DHO plants and 450 SEK/ton plastic waste at CHP plants [29]. By making waste incineration a more expensive option, the idea of the tax is to make recycling of plastic waste (and other material with fossil origin) an economically more favorable option.

9.3.1.2 RESULTS OF THE EVALUATION

The climate tax adds to the costs of the waste-management system. However, the cost-optimizing mix of treatment technologies in the NatWaste model for the year 2030 is not affected by the tax [29]. For the waste fractions where the model can choose between waste incineration and recycling, waste incineration is the cost-efficient technology even with the tax. Note, however, that the optimum mix of technologies in the year 2030 includes no waste incineration with DHO. All incineration has CHP, which means that only the lower tax level is used.

The investigated tax might, however, have some effect in the current system, because some incineration plants today have DHO and would be affected by the higher level of the tax. In addition, NatWaste includes only two average costs for recycling of plastics (from households and industries respectively). Sahlin et al. [40] evaluated the climate tax with a spread-sheet model that included estimated marginal-cost curves for increased rates of recycling of plastic waste from households. This was an attempt to take into account local variations and also variations in the reluctance of households to increase source separation. The results indicate that the climate tax in the current system would increase the recycling of Swedish hard plastics packaging from households by 14%, corresponding to an annual amount of 4 ktonne. Ekvall et al. [48] argue that the effect on the source separation in households in reality might be much smaller, because the climate tax affects the households indirectly only. On the other hand, the tax can contribute to stimulating recycling of materials other than plastics through increased gate fees at waste incinerators or through improved collection systems in general. All this implies that the tax might have some short-term effects on the recycling rate of waste with fossil origin.

The analysis with the NatWaste model is limited to the treatment of Swedish waste. It does not include effects on imports of waste to Swedish incineration plants, an import that has grown in recent years and currently accounts for 15–20% of the total waste incinerated. A climate tax could lead to higher gate fees at waste incineration which in turn, according to Olofsson et al. [51], could reduce the drivers for import of any waste for incineration in Sweden. This could contribute to increasing landfilling, incineration, biological treatment, and/or recycling of various kinds of waste in other countries [48]. However, the tax level required for shifting the cost-effective technology for imported waste has not been analyzed.

All in all, the model results and our analysis indicate that a climate tax on waste incineration in line with the current CO_2 taxation could have a very modest effect on the future waste management system and consequently, the resulting environmental impacts would also be limited.

9.3.2 INCLUDING WASTE IN GREEN CERTIFICATES FOR ELECTRICITY PRODUCTION

9.3.2.1 DESCRIPTION AND ASSUMPTIONS

In the existing system of green certificates for electricity production in Sweden, electricity producers who use renewable sources get certificates from the government. All electricity suppliers are required to have a certain quota of certificates. The suppliers that do not get them by their own production can buy them from other producers. The aim of the system is to increase the production of electricity from renewable sources. The system is further described by Bisaillon et al. [20] and by Bergek and Jacobsson [52].

Currently, electricity production from mixed renewable waste is not included in the system so almost no certificates are given for electricity production from waste incineration. Only the electricity production generated from combustion of separated wood fractions at the incineration plants are included in the current system. In the research program, the following policy change has been studied: certificates are given for the whole mixed fraction of the waste that comes from renewable sources, such as food waste, wood, cardboard etc. but not for the fractions of non-renewables such as fossil-based plastics. The idea of this policy is to stimulate CHP in waste incineration. In the analysis, the price of a green certificate is assumed to be 200 SEK/MWh.

9.3.2.2 RESULTS OF THE EVALUATION

Expanding the system of green certificates to waste incineration will reduce the total net cost of the waste-management system. However, the cost-optimizing mix of treatment technologies in the NatWaste model is unaffected in most scenarios for the year 2030 [29]. The exception is Scenario 2, where the certificate system means that some organic waste fractions are treated by waste incineration instead of anaerobic digestion. The waste incineration increases from 12,820 ktonne (without the policy instrument) to 12,906 ktonne (with the instrument implemented). Electricity production increases slightly, and biogas generation decreases. All this leads to slight changes in the environmental impacts of the system [30].

The modest impact is by large explained by the fact that all waste incineration is CHP in 2030 even without the policy instrument, according to the NatWaste results. The extra incentive provided by the certificates therefore has a limited impact in the year 2030.

9.3.3 COMPULSORY RECYCLING OF RECYCLABLE MATERIALS

9.3.3.1 DESCRIPTION AND ASSUMPTIONS

Although recycling has increased in Sweden, recyclable materials are still being incinerated [5,53]. One policy proposal is therefore to introduce compulsory recycling of recyclable materials, except for materials where incineration leads to lower life cycle environmental impacts. Examples of such materials could be wood waste, yard waste and some types of sludge. A more precise description of the instrument could therefore be *Compulsory recycling of materials defined as recyclable*.

Although this policy proposal is fairly new to the Swedish context, similar policies exist in several places in North America. For instance, in the State of Massachusetts, a ban on incineration and landfill disposal of some recyclable products was introduced in 1990 [54]. The material is only banned from incineration if there are alternative market outlets available. This definition is revised on a continuous basis. Moreover, in the State of Vermont recycling of organic waste will be required [55], and the city of Vancouver has banned disposal of a number of materials including recyclable paper and some containers [56].

9.3.3.2 RESULTS OF THE EVALUATION

Ambell et al. [4] analyzed the cost and environmental impacts of maximizing materials recycling, using the models NatWaste and SWEA. Food waste and other organic waste fractions were not included in this evaluation. In the reference scenario for 2030, the following additional quantities of materials are assumed to be recycled.

- Paper: 1386 ktonne (+90%)
- Metals: 263 ktonne (+14%)
- Plastics: 980 ktonne (+398%)
- Glass: 91 ktonne (+26%)
- Rubber: 63 ktonne (+394%)
- Gypsum: 615 ktonne (+81%)
- Textiles: 205 ktonne (+801%)

These calculations are based on maximum material recovery, meaning all recyclable materials in mixed waste fractions were assumed to be source separated and recycled. A major challenge with this instrument is to decide what material can be recycled. What quantities would be affected in reality depends on the details of the regulations, e.g., on how the concept of recyclable materials is defined, and on practical limitations concerning what can actually be source separated. The SWEA results indicate that introducing a requirement to recycle recyclable materials would lead to significant reductions for the analyzed impact categories: global warming potential, photochemical oxidant formation, terrestrial acidification, freshwater eutrophication, marine eutrophication, and total energy use.

The environmental benefits were valued using the economic valuation method EcoValue 08 [57] showing decreased environmental cost of 1 or 250 billion SEK depending on whether the low or high valuation set is used [58]. It can be noted that all relevant environmental impacts, such as human and eco-toxicological impacts, were not included in the study by Ambell et al. [4] implying that the environmental benefits could be underestimated.

The economic optimization using NatWaste indicates that, compared to the reference case, maximizing recycling increases the overall waste management system cost by 10 billion SEK [4]. This corresponds to an average cost of 3800 SEK/ton of extra waste recycled. This cost is especially sensitive to the data and assumptions regarding plastic waste recycling. In a sensitivity analysis, where the costs of plastic waste recycling is reduced by 33%, the average cost of the obligation decreases to 2400 SEK/ton of extra waste recycled (corresponding to 7 billion SEK). As a comparison, the marginal cost for collection and recycling of plastic packaging has been estimated to be in the order of 4,000 SEK/ton [59]. This marginal running cost is valid for increased collection and recycling of plastic packaging waste in the range of additionally 10–20% of the current collected amounts.

Ambell et al. [4] used constant average unit costs for each fraction, i.e. unit costs did not increase with recycled quantities. It can be expected that in a given time period the recycling cost per ton will increase with higher quantities of materials being separated. On the other hand, the Ambell study does not include any possible future cost reductions from technological developments and from using large scale solutions which may lead to economies of scale and decreased costs with higher recycling rates. The costs of recycling may thus both increase and decrease with increased recycling.

Bernstad et al. [5] did a study which in some aspects is similar to the study by Ambell et al. [4] using the Danish LCA-model EASEWASTE applying it to a residential area in Southern Sweden. They also found that there is a potential for increased recycling and that if this potential is realized, the environmental impacts analyzed would decrease.

Ambell et al. [4] did not take into consideration that compulsory recycling of recyclable materials could contribute to changes in the waste management in other countries. Such effects can be environmentally important and depend on the details of the policy instrument. As discussed above, 15% (about 0.8 million tons) of the waste incinerated in Sweden comprises imported European waste [60], of which the majority, in the short-term, probably would have been directed to landfills abroad. This is valid as long as landfill disposal is the dominating treatment method in the exporting countries. In the coming years, the imports are expected to double [60]. The future import of waste is however very uncertain and depend e.g. on waste policies in the exporting countries. This policy instrument would stop import of recyclable waste to Sweden and possibly, in the short-term, increase landfilling, and, in the long-term, if the European policy of "virtually eliminating landfilling" as referred to in the introduction comes into effect, increase recycling in other European countries. The ban in Massachusetts resulted in some waste being exported to neighboring states for incineration and landfill disposal [54,61].

The social acceptance of this policy instrument was not studied explicitly. However, it seems reasonable that some stakeholders, e.g., waste and energy companies will oppose since it would reduce the quantity of waste available for incineration. Since this policy instrument would require efficient waste separation, at source, after collection or both, high social acceptance is needed for a successful implementation.

9.3.4 TRADABLE-RECYCLING-CREDITS

9.3.4.1 DESCRIPTION AND ASSUMPTIONS

Systems for tradable recycling credits can be designed in different ways. The following version is evaluated within this research program: a minimum recycling level (quota) or rate for a particular material is imposed. To make this happen so-called recycling credits are awarded to the company that use recycled material in the production of new products. The manufacturers of the products containing the material would be required to meet a specific share of recycled material. They could perform the recycling themselves or they could purchase credits from others who have recycled more than their own obligation. A similar system exists in the UK since 1997, and it is known as the Packaging Recovery Notes (PRN) [62,63].

The evaluation of this type of policy instrument is primarily based on relatively simple comparative static models [49], but this is complemented by empirical evidence on the behavior of key secondary material markets such as steel scrap, secondary aluminum, wastepaper etc. (e.g. [41,64]). The specific impacts of tradable recycling credit schemes are likely to be highly context-specific, and thus deserve increased attention in future research.

9.3.4.2 RESULTS OF THE EVALUATION

The impact of this system depends on whether an established market for recyclables exists for the material. It also depends on the geographical scope of the system. A national Swedish system for tradable recycling credits can be ineffective in the case of materials for which an international market for recyclables exists (e.g., metals and paper). There is a risk that although a larger share of the material collected for recycling will be used in Sweden, it may simply increase imports and thus have little or no effect on the total (global) recycling of the material unless the instrument is combined with explicit supply-side policy measures (e.g., waste sorting and collection). Similar trade-related issues have been a concern in the UK system [62]. Important interactions with other policy instruments also need to be addressed. For instance, Matsueda and Nagase [63] show, in the context of the UK scheme, that introducing a tradable recycling credit scheme together with a higher tax at the landfill could in fact raises the amount of landfill waste.

An international system for tradable recycling credits could be more effective. A national system can also work for materials that are mainly traded within Sweden (e.g., glass and gravel). When the system has the same geographical scope as the market, the market impacts of this policy can be described as follows [49]. The implementation of the quota leads to an increased supply of recycled material, and a corresponding fall in the use of virgin material (given a fixed total demand for the material). The price faced by virgin material suppliers will therefore fall. In order for the quota obligation to be fulfilled the suppliers of secondary materials will receive extra revenue per unit material supplied. The producers of material-containing products can in turn finance their purchases of recycling credits by levying an extra fee on end consumers.

The conclusion is that a tradable recycling credit scheme has a potential environmental gain, at least when the geographical scope of the system is at least as large as the geographical scope of the market. With a well-functioning market for certificates or credits (i.e., many actors, low transaction costs etc.), cost-efficient solutions for recycling are sought. This means that a given recycling level can be achieved at minimum cost.

However, when the supplied quantity of secondary, recycled material is forced to increase, the cost to achieve the goal will be uncertain. The scheme will increase the price of secondary materials, but the supply of secondary materials is typically not very sensitive to changes in the price (e.g., [41]). This means that the price of the material and, hence of the recycling credits, might have to be very high to reach the target level of recycling. The cost is also expected to rise steeply if the target is increased, e.g., if the practical difficulties in the recycling process have been underestimated. Still, the environmental benefits of increased recycling are not likely to rise steeply in a similar way. For this reason, it can be argued that the economic efficiency of introducing this type of quantity-setting measure may be low (based on Weitzman's [65] seminal studies on the choice between price- versus quantity-based policies). It might instead be more efficient to stimulate recycling through the use of price-based policy instruments (e.g., virgin material taxes; see, however, Section 3.8).

The risks for high compliance costs can be alleviated if the system of tradable recycling credit is combined with measures to increase the collection of used materials for recycling. Exceedingly high costs can be completely avoided by allowing producers to stay outside the system for a fixed fee per ton of material. This fee will then set a ceiling for the price of recycling certificates. In a well-functioning market for certificates, the companies have the freedom to choose and flexibility to develop and search solutions for cost-efficient recycling. This would probably have a positive influence also on the producers′ acceptance for such a system.

On the other hand, this kind of system could also neutralize the effect of voluntary efforts to increase recycling levels. This is because the quota puts a cap on the amount of recycling, and voluntary initiatives will not add to this cap; instead they will simply make it easier and thus cheaper for product manufacturers to comply with the cap. From the perspective of the "volunteers" (e.g., consumers with strong preferences for recycled products), the acceptance can therefore be low.

9.3.5 WEIGHT-BASED TAX ON INCINERATION OF WASTE

9.3.5.1 DESCRIPTION AND ASSUMPTIONS

This policy instrument is a weight-based tax on incineration of solid waste. Incineration of waste from both renewable and non-renewable materials would be taxed. Different versions of the tax could be implemented. The tax could be introduced for only household waste or for all types of waste.

The tax could also be introduced with a tax reduction for plants with combined heat and power production.

Slightly different versions of taxes on incineration of waste have been evaluated. Björklund and Finnveden [66] studied the environmental impacts using an LCA model of a tax of 400 SEK/ton on incineration of household waste without tax reduction for CHP. Sahlin et al. [40] studied the incineration tax that was implemented in Sweden from the year 2007 to 2010 and compared it to the net marginal costs of waste treatment alternatives. This tax was slightly higher than 400 SEK/ton for household waste in DHO, but much lower for CHP. This construction aimed to stimulate CHP and also to mimic the tax on fossil fuel used in the Swedish district-heating sector based on average contents of fossil material in municipal solid waste. In both cases the tax was assumed to have an impact on the gate fee for waste incineration. This makes waste incineration less economically competitive in general compared to alternative treatment such as material recycling and biological treatment.

9.3.5.2 RESULTS OF THE EVALUATION

Increased gate fees will affect all actors that deliver household waste at the incineration plant; waste collection companies and similar. Their cost increase is likely to be transferred to the households and the companies, and increase their cost for waste treatment.

The proposed design of the tax is expected to increase recycling only to a small extent, and give rise to small environmental improvements and energy savings [40,66]. Using an optimizing spreadsheet model (cf. Section 3.1.2), Sahlin et al. [40] predicted the largest effect on household waste to be on biological treatment of kitchen and garden waste, which would increase from 16 to 17% (level of 2006) out of the total treatment of household waste. Ekvall et al. [48] argue that even this modest result might be an overestimate. In order to have an effect on the treatment of household waste, the tax must affect the source separation in households, and an incineration tax affects the households only indirectly.

If the tax includes a reduction for CHP, waste may be redirected to CHP plants from heat-only boilers. This is expected to give further environmental improvements, at least in the short-term when not all waste incineration has CHP. On the other hand, the tax reduction will of course also lower the economic incentive for finding alternative waste treatment methods.

Concerning the acceptability of this policy instrument it can be noted that the waste incineration tax that was introduced in 2007 met strong resistance from several stakeholders including the municipal waste management companies although it had support from other stakeholders, including recycling companies [67]. After a general election and change of government, the tax was eventually removed, also indicating different political opinions concerning the tax.

9.3.6 WEIGHT-BASED WASTE COLLECTION FEE

9.3.6.1 DESCRIPTION AND ASSUMPTIONS

The idea of a weight-based fee is that the households pay per mass of waste discarded. The weight-based waste collection fee can have an effect in two ways:

- an economic incentive to reduce the quantity of residual waste though prevention, recycling, or irregular or illegal waste treatment, and
- raised attention to waste-management issues that, at least temporarily, can result in waste prevention and increased recycling.

Bisaillon et al. [20] propose to assess a waste collection fee for households with a fixed part (850 SEK/household and year) and a variable part (2.12 SEK/kg residual waste). Based on earlier studies [68,69] it is assumed that this leads to a 20% reduction of the collected residual waste. Since there are several plausible explanations to the reduction we have analyzed three extreme alternatives [29,48]:

1. All reduction in residual waste is due to prevention of waste with the same composition as the average residual waste.

2. All reduction in residual waste is due to an increase in source separation for home composting (50%) and materials recycling (50%).
3. All reduction in residual waste is due to illegal treatment: e.g. burning of combustible waste in private stoves or dumping of food and garden waste in the forest.

9.3.6.2 RESULTS OF THE EVALUATION

From an environmental perspective, the fate of the waste that is not collected as mixed residual waste is important. Arushanyan et al. [30] show that Alternatives 1 and 2 could lead to environmental benefits. The waste prevention in Alternative 1 could reduce greenhouse gas (GHG) emissions in the year 2030 by 2300 kton CO_2-eq. Using two versions of the Ecovalue method, Arushanyan et al. [30] calculated the total environmental benefit from the policy instrument to correspond to 1 or 128 billion SEK for the two sets of values in the method.

The increased recycling in Alternative 2 could reduce GHG emissions by 600 kton CO_2-eq. The total environmental benefit was calculated to 0.2 or 1.3 billion SEK [30].

The environmental impacts of Alternative 3 were not evaluated in this study. However it is clear from previous studies that uncontrolled burning of waste can lead to significant emissions of hazardous compounds [70]. Therefore emissions from uncontrolled burning can be significant compared to the total emissions, even if the amount combusted in uncontrolled burning is just a fraction of a percent (ibid).

The risk of increased uncontrolled burning and other illegal treatment is lower when the households are driven by strong pro-environmental attitudes, and higher when they are simply interested in the impacts on the household budget. After deep-interviewing 42 households in Gothenburg, where the fee was recently introduced, Schmidt et al. [71] concluded that the main driver for change is not the fee as such since it is small compared to the total household budget. Instead, the households seem to be affected mainly by the norm-activating information that was distributed as the fee was introduced and by the regular feedback from the system, which both confirm the feeling that "sorting is doing the right thing". This is also in line with the results of Sterner and Bartelings [72] that economic incentives are not the only driving force behind a reduction in waste. This implies that the increase in uncontrolled burning could be small.

The consequences in the year 2030 depend on how the society and associated norms develops in the future. Ekvall et al. [48] argue that the weight-based waste fee can be expected to have good environmental consequences in scenarios where the environmental awareness is great (Scenarios 1 and 4; [22]). In scenarios where private economic impacts are among the dominating driving forces (Scenarios 2 and 3), the risk of a significant increase in illegal treatment is greater. Fullerton and Kinnaman [73] as well as Walls and Palmer [74] show that if illegal dumping behavior is present a combined output tax and recycling subsidy could be an efficient second-best policy. The tax discourages production of waste-intensive products, while the subsidy encourages substitution of secondary materials for virgin materials.

The connection between external scenarios and the effect of a weight-based fee might, however, be more complex than this. Andersson et al. [44] suggest that the policy instrument, since it is a market-based instrument, will most likely function well in market-oriented scenarios (Scenarios 2 and 3) where the individual takes a large responsibility, and that it will not be as effective in the "sustainability-scenarios" (Scenarios 1 and 4) where sustainability is a natural part of the society.

A weight-based fee requires technological and administrative systems: trucks with scales, etc. The associated costs are likely to differ across regions, thus suggesting that it should not be implemented uniformly across the entire country. It is in general well accepted by the households [71]. An exception might be households with seemingly unavoidable large volumes of residual waste [44,71], for example families with children in diapers.

The legitimacy of a weight-based waste collection fee might, however, decline over time due to, for example, distrust in the system or if the recycling stations are not emptied often enough to absorb the increased flow of source-separated materials. In order to have beneficial long-term effects the fee should be complemented with increased collection frequency at recycling stations and kerbside containers. It should also be complemented by norm-activating information [75]. Such information could strengthen the effect since the information will underline the economic incentive. A trust in the environmental effectiveness of the system is an important determinant for the attitudes towards recycling schemes [76].

9.3.7 DEVELOPED RECYCLING SYSTEMS

9.3.7.1 DESCRIPTION AND ASSUMPTIONS

Source separation can be negatively affected by practical aspects as well as uncertainty among people [43]. The collection system can be improved to make things easier for the households, for example through kerbside collection (reduced transport distance for the household) or by a collection system based on material streams (e.g., plastics) instead of product groups (e.g., packaging). The policy instrument evaluated in this study represented a combination of these, property-close or kerbside collection of material streams [20].

9.3.7.2 RESULTS OF THE EVALUATION

Hage et al. [69] provide an econometric analysis of the collection of plastic packaging waste across almost all Swedish municipalities, and show that the presence of kerbside recycling and the number of drop-off stations per square kilometer, respectively, have significant impacts on the reported collection rates. Also Söderholm [75] emphasize the relation between increased availability of recycling opportunities for the households and increased collected amounts of recyclables. This indicates an overall important potential to increase collection by improving the collection infrastructure.

Kerbside collection increases the costs of collection and transportation for the waste management company, and this can be relatively high in sparsely populated regions (e.g., Kinnaman [77]). On the other hand, the transport needs for the households decreases. Previous studies indicate that the frequency of travels for the sole purpose of dropping off waste is often relatively high [75]. If households in general do not combine travels for drop off of packaging waste with other travels, overall transportation costs could actually be reduced through the introduction of kerbside collection. This is because the introduction of kerbside collection means that the uncoordinated trips of households to the recycling stations are displaced by centralized transport pick services.

Another way of developing the collection system is to collect waste in material streams instead of the current Swedish system where only packaging materials and paper are collected. A pilot test where plastic and metals from households were collected was organized by the Swedish EPA [78]. The results indicate that the system would be easier for households to understand and that their motivation would increase and therefore also the collection of recyclable materials. This would have environmental benefits. In order to ensure that the collected materials would be recyclable, further measures may, however, be necessary [78].

Andersson et al. [44] draw the overall conclusion that developed collection systems (including property-close collection and/or collection in material streams) can be an effective way of increasing the collection from households. This, however, requires that systems are adapted to the needs of the households, the knowledge and motivation among households are increased, and the number of fractions that are sorted at home should preferably not increase.

Based on the evaluations made, developed collection systems would likely contribute to increase recycling and positive environmental effects but the magnitudes of these effects are uncertain. The advantages are that the customer may see it as a higher level of service and that the facilitated collection may increase the amount of waste collected [44]. However there are also some drawbacks like increased needs for heavy transport (on the other hand the household's transports of waste will decrease).

9.3.8 TAX ON VIRGIN RAW MATERIALS

9.3.8.1 DESCRIPTION AND ASSUMPTIONS

In order to reach an efficient use of raw materials, taxes should be introduced if there are significant external costs associated with raw materials extraction or use. Since there are environmental impacts associated with extraction of raw materials that are not internalized, a raw material tax can increase the economic efficiency and reduce environmental impacts. Moreover, if the market actors are using a higher discount rate (rate-of-return requirement) than what is socially optimal, too much material could be extracted. In order to change this, a raw material tax can increase the efficiency. Raw material taxes can also be used as a second best option to reduce environmental impacts further down in the product chain [50].

Taxes on raw materials can be designed in a number of different ways [50]. They can be broad taxes covering a large number of materials or more specific taxes for selected materials. Sweden already has a tax on natural gravel and an energy tax on fossil fuels. In this program we have evaluated two broad raw materials tax proposals, both described by Bisaillon et al. [20]:

- A 10 SEK/ton tax on non-renewable materials (excluding fossil raw materials and plastics) extracted or imported and then used in Sweden.
- A tax on all fossil raw materials similar to the one currently applied on household heating oil (3804 SEK /m^3) and an associated 5000 SEK/ton tax on imported plastics.

Forsfält [27] analyzed the impacts of each of the two taxes separately using the EMEC model. The assessment of the first tax was limited to a test on how a tax on the mining of metals in Sweden affects the economy and waste generation. Adjustments were made so that exports were not taxed while imports should be taxed similarly to domestic production. No attempts were made to analyze how the economy is affected by a tax on raw materials in refined products. The analysis with EMEC was complemented by more conceptual analyses [48] as well as a synthesis of the empirical experiences of aggregates taxes in Europe [50].

9.3.8.2 RESULTS OF THE EVALUATION

The EMEC results on the 10 SEK/ton tax on metals and mining waste indicate that the impact on the total amount of waste is relatively small:

the demand for products from the mining industry decreases somewhat [27]. Sectors "downstream" are affected since the relative price of ore increases; how much depends on companies' ability to substitute between different inputs. For the iron and steel industry and other metal industries, the possibilities for substitution are assumed to be limited and therefore the value added falls in these sectors. Production levels in other sectors of the economy are almost unaffected, as is the total GDP and household income. The total quantity of non-hazardous wastes in the EMEC model for the year 2030 is reduced by 30–40 ktonnes (depending on scenario). The quantity of hazardous waste is reduced by 4–6 ktonnes.

When interpreting the results it should be noted that only some of the possible impacts are included in the model. The possibilities for switching between metals and other materials are possible, but the change from virgin metal products to recycled metal products is not modeled in EMEC. Since this is a change that can be expected, and a change that could also alter the waste amounts, the possible reduction in waste amounts in Sweden may be underestimated. In addition, only impacts in Sweden are included in the model.

The results of the model experiments concerning the second raw-materials tax, 5,000 SEK/ton plastics were different in structure, this since all production sectors in the economy are affected [27]. However, the model results suggest that the net effect on waste generation is still small. The tax actually increases the quantity of non-hazardous waste in the scenario "Global markets". In other scenarios, the quantity is reduced by 7–13 ktonnes. The quantity of hazardous waste is reduced by 4-6 ktonnes in the model, again depending on the scenario.

The small effects on the waste flows means there are almost no changes in terms of environmental impacts: each of the two taxes results in less than 1% of improvement in each impact category according to the results from the LCA model SWEA [30].

These results are consistent with empirical experiences of virgin materials taxes in Europe [50], which suggest that they may have little effect on the use of recycled materials unless they are complemented by measures to promote waste sorting activities.

In the long term, Forsfält [27] argues, a combination of taxes on specific raw materials and broad taxes are probably most effective. Söderholm [50] in turn argues, in line with Walls and Palmer [74], that the policy instrument can preferably be combined with policy measures that promote increased waste sorting (e.g., at construction and demolition sites). Walls and Palmer [74] find that no single tax can generate the optimum level of both downstream and upstream waste disposals and that multiple policy instruments are necessary to fully internalize the externalities. These results are also empirically illustrated in Palmer et al. [79]. The material tax in itself can be expected to have moderate effect because own-price elasticity of demand is often low for materials. The effect on recycling is likely to be even lower, because the supply of recyclable material is also not very sensitive to changes in price. The effects will however depend also on the level of the taxes.

The development of raw material taxes requires some additional work before they can be implemented. Especially issues related to imports and exports may be problematic and need some further considerations. According to Ekvall et al. [48] the effect on the competitiveness of the Swedish industry will be much smaller if the tax is not calculated based on the quantity of materials extracted or imported to Sweden, but on the estimated quantity of non-renewable raw materials that are extracted to produce products that are used in Sweden. This might allow for introducing raw materials taxes that are much higher than the 10 SEK/ton investigated here. Such a tax could have a much greater impact on the material efficiency of the economy and, hence on the waste quantities.

It should be noted that a high tax on virgin raw materials is ineffective if it mainly means that products based on virgin raw materials are used in other countries, while Swedish citizens use products based on recycled materials. In this situation, the tax affects the transportation patterns of materials and products, but it does not significantly affect the material efficiency or the global recycling rate. This ineffectiveness is comparable to the lack of effects of a national system of tradable recycling certificates for materials with a well-established, international recycling market (see Section 3.4).

9.3.9 ADVERTISEMENTS ON REQUEST ONLY

9.3.9.1 DESCRIPTION AND ASSUMPTIONS

This policy instrument prevents companies from direct advertising to mailboxes without a posted sign saying yes to advertising (the opposite from the present situation when household have to say no to such commercial advertising) [20].

9.3.9.2 RESULTS OF THE EVALUATION

"Advertisements on request only" have the potential to reduce the annual wastepaper quantity by up to 12 kg/person and year with current levels of paper consumption [20]. For Sweden this corresponds to 0.11 Mtonne/year (ibid.). This amounts to about a 20% reduction in the waste intensity coefficients for paper waste from households [42]. The reduction in paper waste is expected to result in a corresponding reduction in environmental impacts from the production of the paper leading to overall reduced environmental impacts [80]. However, part of the effect may be offset by more paper being used for other purposes and/or in other countries and that it can increase the production of other types of advertisements [42].

From a psychological and ethnological perspective, Andersson et al. [44] assess this policy instrument to be both effective and legitimate. It makes the environmentally friendly alternative the default alternative, thus contributing to an overall alignment of values and policies. A disadvantage may be that for some groups it could potentially be more difficult to get access to some local and commercial information.

9.3.10 DIFFERENTIATED VAT

9.3.10.1 DESCRIPTION AND ASSUMPTIONS

A differentiated value added tax (VAT) could aim at shifting households' consumption from goods to services and to eco-labeled goods [20]. This could make the consumption more environmentally friendly and less material-intensive.

Using the EMEC model, Forsfält [27] studied the effects of such a differentiated VAT on the Swedish economy and on future waste quantities. The proposal evaluated by Forsfält [27] is that the VAT on all households' consumption of services except transportation should be equal to the currently lowest VAT rate, which implies a reduction from the present 25 (or 12)% to 6%.

9.3.10.2 RESULTS OF THE EVALUATION

The EMEC results depend on the way the tax cut is financed. If it is financed by a decrease in government transfers to households, the consumption of services in the year 2030 increases by 3.6% and households' waste generation falls by about 1 percent, compared to a scenario without this policy instrument [27]. The consumption of goods falls but total consumption expenditures are almost unchanged. Investments fall marginally as do imports and exports. The GDP decreases by 0.1%. Production increases in the service sectors at the expense of commodity producers, but the relative increase in the service sector's value added (0.3%) is less than the increase in service consumption (note that the service industries production is considerably larger than households' consumption of services).

If the tax cut is instead financed by increasing the VAT on goods, then the change in the consumer price of services relative to that of goods is greater, something which in turn causes waste generation to fall by about 1.5% [27]. Also in this case, GDP decreases by about 0.1%.

The results thus show that the waste is reduced more than GDP due to the VAT change. It is the redistribution of consumption that provides the reduced amount of waste. Forsfält [27] concludes that differentiated VAT provides a change in a more sustainable direction although the model results indicate that the resulting waste reduction is modest.

The environmental impact assessment was performed on the case where tax cut is financed by a decrease in government transfers to households, resulting in a waste reduction by 125 ktons. The results indicate environmental gains mainly through avoided production of food, textiles, and steel, according to results from the SWEA model. The expected improvement varies across the impact categories, in the range of 1–7%. For climate change the environmental gain in SWEA is 5% [30]. It is interesting to note that the reduction of environmental impacts is larger than the reduction in waste volume. This is an indication of the importance of waste prevention. Households' real incomes are not affected significantly by the investigated change in VAT, which means that households' total consumption capacity only decreases marginally [27].

In conclusion, the introduction of a differentiated VAT seems rather unproblematic according to these assessments and gives a change in a more sustainable direction. The indicated waste reduction is small but a combination with other instruments, such as information, could improve efficiency.

9.3.11 ENVIRONMENTALLY DIFFERENTIATED WASTE FEES

9.3.11.1 DESCRIPTION AND ASSUMPTIONS

An environmentally differentiated waste fee would provide lower fees for households and companies that sort out more fractions. The idea behind environmentally differentiated waste fees is to achieve a stronger link between household waste fees and the environmental impact of the waste treatment [20]. This type of policy measure is already in use in some Swedish municipalities, e.g., when households separate food waste for biological treatment and then get a lower waste fee (ibid).

In our evaluation, it is assumed that the introduction of the policy measure would give incentives for 20% of the households and firms to sort in three fractions (food waste, packaging and combustible) and that these actors sort out 60% of their food waste and 50% of the packaging materials.

9.3.11.2 RESULTS OF THE EVALUATION

One benefit of this policy instrument is that it provides a clearer connection between the fees and the environmental impacts of waste treatment [44]. It will contribute to an alignment of environmental impacts, policy and norms. However, there may be a risk for fraud and dumping in order to get lower fees in the same way as for weight-based waste fees. This instrument should thus be complemented with norm-activating information. Andersson et al. [44] suggest that environmentally differentiated waste fees should also be combined with improved collection systems. Then the households would face less barriers as well as stronger incentives to sort more efficiently and the fractions could be cleaner. The environmental benefits of this policy instrument can be assumed to be similar to a weight-based waste-collection fee that results in increased source separation and recycling (see Section 3.6).

9.3.12 IMPROVED INFORMATION

9.3.12.1 DESCRIPTION AND ASSUMPTIONS

The evaluation includes both information to households and information to companies and organizations. Bisaillon et al. [20] mention two principal information categories:

- Procedural information, telling for example how, where, and when people should hand over source separated or non-separated waste.
- Declarative or norm-activating information, aiming at giving people motivation why they should source separate and the effects of source separation.

In the analysis, it has been assumed that information is combined with other policy instruments when they are implemented.

9.3.12.2 RESULTS OF THE EVALUATION

The effects of information on waste flows and waste treatment are difficult to quantify. Ekvall et al. [42] suggest that an effective and persistent information campaign could reduce the waste quantity from households in the year 2030 by 10%. They argue that the effect on waste from companies is likely to be smaller. However, these effects will depend very much on the details and context of the information.

Andersson et al. [44] assess the acceptance of information as a policy instrument. This assessment covers both information to households and information to companies, and includes a discussion of how the information should be designed to provide the desired effects. For example:

- Information should be combined with other instruments, for example to make way for other policy instruments.
- Information should be adjusted for each recipient group.
- Information can be given in the form of feedback.
- Information should be conveyed by a credible source.
- Information should rely on ethical norms of what people should do.
- In Sweden, the environmental awareness is generally high. Declarative information is then less important, but it is necessary for groups with low environmental awareness.

One difficulty with information is to reach the recipients, because of the high total information pressure on people. The message is often not noticed if it is not relevant for the recipient at the moment it is received. Another difficulty is that paying attention to the information is voluntary. People who are not interested in waste can simply neglect any information provided.

Andersson et al. [44] still conclude that information is important for the future waste-management, especially in scenarios with a high degree of environmental awareness. Information can be an effective policy instrument in several ways. As a separate instrument it can be used for both affecting the behavior and giving procedural information. It is a necessity when implementing other policy instruments, for example when paving the way for the instruments. The effect of information can also be catalyzed in combination with other instruments. This is also in line with the results of Brekke et al [81] indicating that perceived responsibility is a determinant for reported recycling behavior, but also that uncertainty in the information for example about other people's behavior could cause reluctance to accept responsibility. Also Bruvoll and Nyborg [82] conclude that people are willing to conform to social norms, even if it comes with perceived costs in terms of time or work.

9.3.13 MANDATORY LABELING OF GOODS CONTAINING HAZARDOUS SUBSTANCES

9.3.13.1. DESCRIPTION AND ASSUMPTIONS

Bisaillon et al. [20] describe this policy instrument as a requirement to label all goods containing at least 0.1% substances with very high acute toxicity, allergenic, high chronic toxicity, mutagenicity, or other hazardous properties. Examples of goods that could be labeled are, for example, furniture, shoes, clothes, toys, electronic devices and other goods that people usually do not associate with hazardous content. Chemicals, pesticides, solvents and similar are not concerned since they already are included in other labeling rules and regulations.

9.3.13.2 RESULTS OF THE EVALUATION

Ekvall et al. [42] suggest that mandatory labeling can affect consumer choices and reduce the use of such products. It would thus contribute to waste prevention by minimizing the hazardousness of waste. Andersson et al. [44] mention several advantages with this policy instruments:

- The consumers themselves decide if they want to contribute to a better environment when shopping.
- If there are alternatives, the consumers will have the possibility to avoid products containing hazardous substances.
- Consumers without an active interest in environmental issues can still be prone to avoid exposing themselves and their family to hazardous substances. This is likely to make negative environmental labeling (of hazardous substances) more effective than positive environmental labeling (eco-labeling).

Among the drawbacks the following can be noted [44]:

- The system must be mandatory. A voluntary system can lead to confusion if a lot of products and trademarks are not included.
- It is easier for a manufacturer to accept an eco-labeling system.
- There are examples when the labeling can lead to confusion, e.g., if a product has both positive and negative environmental properties.

All together Andersson et al. [44] conclude that labeling of environmental hazardous substances is likely to be an effective policy instrument to decrease the content of hazardous substances in the waste.

9.4 DISCUSSION

The present study is fairly unique in its scope. Many of the existing policy instruments and also much of the scientific literature tend to focus on specific waste types such as packaging [83,84] or WEEE [85,86,87], or on municipal waste. This study instead takes on a broader perspective and has addressed the whole national waste management system and its stakeholders. Therefore it deals with many different policy instruments, which are evaluated from several perspectives, and that in turn can be implemented at various parts of the system and for many types of waste. This makes it possible to identify policy combinations and compare different instruments and policy mixes, both existing and not yet implemented ones. Based on the learnings from our and other assessments some conclusions can be drawn on the implications for Swedish waste management policy, which policy instruments that have potential, what needs to be developed further, and parts of the system where more and innovative policy instruments are needed.

9.4.1. DEVELOPMENT OF POLICY INSTRUMENTS TOWARDS IMPLEMENTATION

The evaluations summarized above show that some policy instruments have the potential to contribute to more sustainable waste management, whereas other show a low potential or need for redesign. This section includes brief discussions on the instruments that showed potential in the evaluations and with suggestions for adjustments and redesign.

9.4.1.1. COMPULSORY RECYCLING OF RECYCLABLE MATERIALS

Compulsory recycling of recyclable material can lead to a high environmental gain compared to the potential environmental impacts of the business as usual scenario (c.f. Ambell et al. [4]). This is therefore a policy instrument that should be further investigated. There are financial costs for increasing material recycling, but there are significant environmental gains to be made. A number of aspects need to be further studied. Which materials or products should be included and to what extent? Should there be an existing market for the recycled material or is it enough if it can be expected to develop? What are the costs for recycling of individual materials and what are the environmental benefits? What are the implications if international waste trade is considered as well?

9.4.1.2. WEIGHT-BASED WASTE FEE IN COMBINATION WITH INFORMATION AND DEVELOPED COLLECTION SYSTEMS

A weight-based fee is already used in several Swedish municipalities and would be relatively easy to introduce in others. Although experience indicates reduced collection of waste, the reasons for the reduced waste and thus the potential environmental impacts are uncertain. However, in combination with other policy measures it might have a larger effect and Schmidt et al. [71] found that it seems to function more as an information instrument than as a financial incentive as it gives positive feedback on the right waste handling practice. Since there might be a risk of illegal waste treatment involved, it could be a good strategy to combine this policy instrument with an improved recycling system and increased information on the negative consequences of, for example, uncontrolled burning.

9.4.1.3. MANDATORY LABELING OF PRODUCTS CONTAINING HAZARDOUS CHEMICALS

An important part of waste prevention is to minimize the content of hazardous substances in the waste. Waste policy is therefore linked to chemicals policy. Labeling of products containing hazardous chemicals can be a useful instrument. However, it probably needs to be implemented at the international level for example within the European Union. Sweden can take a role in developing such an initiative.

9.4.1.4. ADVERTISEMENTS ON REQUEST ONLY AND OTHER WASTE MINIMIZATION MEASURES

Advertisements on request only can be efficient as a policy instrument since it is expected to have an impact on the reduction of paper waste, and most people are expected to accept. This is an example of a waste minimization policy measure. There are others, for example supporting re-use of products by making it easier for people and organizations to sell or give products to charity before they become waste, to reduce food waste, and to reduce beverage packaging [88,89,90,91]. Each one of these waste minimization policy measure tends to have a rather limited impact in terms of reduced waste amounts. However, if waste minimization also leads to reduced production, the environmental gains can be larger than the reduced waste amounts indicate. Several measures together can therefore make significant contributions.

9.4.1.5. DIFFERENTIATED VAT AND SUBSIDIES FOR SOME SERVICES

According to our assessments, a differentiated VAT where the tax is reduced for services is expected to be beneficial in terms of environmental impact due to a redistribution of consumption. The households' real incomes are not affected significantly. The indicated waste reduction is small but a combination with other instruments, such as information, could improve efficiency.

In order to achieve more significant impacts it would be interesting to test a more extended VAT differentiation. In Sweden the standard rate today is 25%, but this could possibly be increased while VAT on households' consumption of services could be decreased even more compared to the tax changes that have been evaluated. Changes in VAT would have other macroeconomic effects, such as impacts on employment. A differentiated VAT could thus be of interest in a broader tax reform.

This policy instrument could also be made even sharper by not only reducing VAT but also introducing subsidies to certain services such as leasing, repairing, renovation and second hand in parallel to current subsidies on domestic services.

9.4.1.6. INFORMATION

Information is a necessary policy instrument in order to successfully implement most other policy instruments. It is necessary but not sufficient in itself. In order to be efficient, information should be combined with other tools and designed in relation to the specific situation

9.4.2. NEW IDEAS FOR POLICY INSTRUMENTS

Besides the policy instruments evaluated and discussed above, there are several other instruments that are of interest for further development. Some of these are discussed below as well as some instruments that were evaluated but need modifications.

9.4.2.1. A GENERAL (RAW) MATERIALS TAX

Taxes on the use of raw materials are likely to be necessary in a more sustainable society. They can be designed in a number of different ways and further work in this area is needed. A possible development of the tax on non-renewable raw materials discussed above is a materials tax where the tax is the same for virgin and for recycled material. This policy instrument could serve to increase material efficiency all through the life cycle since materials will become more expensive. If a high materials tax is introduced in Sweden, however, measures must be taken not to hamper the competitiveness of the Swedish industry. Such a measure could be to extract the tax on materials and products used in Sweden only, and on imported products as well as on domestically produced products.

9.4.2.2 RE-USE CERTIFICATES

As a further development of recycling certificates, it is possible to conceive a system of reuse certificates. This would stipulate that a specific share of a specific type of products or components, for example bricks or roof tiles in new buildings, should be reused. Certificates would be awarded to construction companies that, to continue the example, reuse bricks or roof tiles. If a construction company has a lower share of reused bricks than stipulated by the policy instrument, they would need to buy certificates from companies that reuse more than they need. This policy instrument could serve to create markets for reused components and products.

9.4.2.3 BRINGING DOWN TRANSACTION COSTS

There may be high search and transaction costs associated with recyclable materials related to incomplete information [92]. Users and suppliers of recycled materials can have problems finding each other. There can be a lack of information about the quality and properties of potentially recyclable or reusable materials and products. In addition, the information may be asymmetric so that the supplier knows more about the material or product than the potential buyer. A broad range of policy instruments can be used to support the markets in these situations. Examples include support for establishing market places, information hubs or hiring waste brokers. It can also be useful to support the establishment of different certification schemes and quality standards. Also requirements to provide information on the content of different materials, e.g. in products or building materials, may be useful. Users may lack knowledge about how to use recycled materials, for example in the production of new products. In such situations, information and education of users, e.g. product designers, could be useful. Other examples include supporting industrial symbiosis by removing institutional barriers for increased recycling of industrial by-products and wastes (e.g., Watkins et al. [93]). Many of these initiatives exist already today, but there is a need to investigate and develop these instruments further. Many waste management policy studies in the environmental economics literature (e.g., [73,74]) assume implicitly the presence of efficient private markets for recycled materials. Only a few analyzes the case of second-best optimums assuming that recycling markets are not operating (e.g., [94]), while Calcott and Walls [95] represent one of few studies addressing the presence of existing—but imperfect —recycling markets.

9.4.2.4 REQUIREMENT OF DESIGN FOR RECYCLING

Another type of barrier may be the existence of technological externalities [92]. An example of such an externality is if the production of a product is made in such a way that the cost of recycling is increased, but neither the producer nor the buyer of the product has to pay for this extra cost. A concrete example may be the current design of dishwashers which often makes it difficult and costly to dismantle it in a way that the copper can be separated from the steel, making the recycling of both the copper and the steel less effective [96]. If this extra cost for dismantling is not covered by the buyer or seller of the dishwasher, an external cost is passed on.

Calcott and Walls [95] note that in practice it is difficult for policy-makers to attain a first-best outcome in the case of these design issues, primarily since product-specific taxes that vary with the degree of recyclability are difficult (and costly) to implement. Strong incentives for recycling design require, though, that recyclers keep track of exactly which firms' products they are recycling. Extended producer responsibility, if the responsibility is individual for the producing companies, may be a way of moving the extra costs of recycling to the producer and thus providing an incentive for a better design. If the extended producer responsibility is collective, this incentive will however not exit, and the technological externality will still be there. A requirement for design for recycling may in this case be more effective. This could possibly be introduced in the Ecodesign directive [20].

9.4.2.5. TAX ON HAZARDOUS SUBSTANCES

Very hazardous substances should be banned from use. There will however always be substances that society cannot or does not want to ban completely, but still want to minimize the use of. Taxation on the use of hazardous substances can be made in different ways [97]. Taxes can, for example, be put on specific substances. The idea is that the use of these substances will decrease and, eventually the waste may contain less of the hazardous substance and become less hazardous. Taxes can also be placed on all substances fulfilling certain criteria. Different options for the taxation of hazardous substances should be further developed and evaluated.

9.4.2.6. DEPOSIT AND REFUND SYSTEMS

Deposit and refund systems where the buyer of a product pays a deposit which is paid back when the product is left for waste treatment can be an effective instrument for making sure that products are collected in a proper way. This is a type of instrument to be used for product groups of special importance. In some cases, a refund, or a bonus, could be paid without actually having collected a deposit, to make sure that products are collected. This could be relevant for products for which the environmental impacts would be especially worrisome if the waste products are not properly handled.

9.4.2.7 A BROADER LANDFILL TAX

Many waste materials are currently exempt from the landfill tax in Sweden. This is the case for waste fractions where there are no alternative treatments. With such exempts there are, however, little incentives for developing new waste treatment methods or for waste minimization. It could therefore be useful to introduce a lower landfill tax also for these waste materials. This tax should preferably correspond to the environmental costs of the landfilled waste, and is endorsed in previous studies. For instance, Kinnaman [77] argues that landfill taxes are typically inexpensive to administer and unlikely to cause illegal dumping. In many ways it may reduce the need for other initiatives, such as weight-based fees and kerbside collection

9.4.2.8 REQUIRED STORING OF PLASTICS THAT CANNOT BE RECYCLED

In most cases, recycling of plastics is the most environmentally friendly option (e.g., [98,99,100]). However, if recycling is not possible, storing or landfilling of plastics can produce less emissions of gases contributing to climate change compared to incineration, at least in the time perspective of a century or shorter, even if the heat and electricity production from the incineration is credited by assuming that it replaces energy sources typically used in Swedish and European conditions [101]). Storing of plastics is however currently not allowed because of the landfill ban on organic and combustible materials. A change in this policy could be one of the most effective ways of reducing the emissions of greenhouse gases from the waste sector.

9.4.2.9 INCREASED CONTROL AND MONITORING BY AUTHORITIES

This policy instrument means that more resources are allocated to monitoring and control of industry and other commercial activities, particularly regarding the waste management [20]. The control can be both within the existing areas of control or more specific on for example waste prevention. The idea is that increased control by the authorities would lead to an increased awareness regarding wastes which would lead to a decrease of hazardous waste in mixed waste and increased sorting in general. This might in turn lead to, for example, more waste being recycled and decreased amounts of hazardous waste.

9.4.2.10 WASTE MINIMIZATION IN ENTERPRISES (INDUSTRIAL WASTE-PLAN REQUIREMENTS)

Bisaillon et al. [20] describe several alternative policy instruments for waste minimization in enterprises. One is to require a waste plan, including plans for waste minimization, from companies generating more than 2000 tons/year non-hazardous waste or more than 2 tons/year hazardous waste. These figures are in accordance with the requirements in the Environmental Code to submit an annual environmental report. Another alternative is to require detailed descriptions of activities to minimize waste when applying for environmental permits and in the annual environmental reports require a follow-up of these actions and continuous up-dating of the minimization plan. For all of these alternatives the policy instrument aims at creating conscious and permanent deliberations on waste prevention and recycling in the companies.

9.4.3 CHOOSING AND COMBINING POLICY INSTRUMENTS

One important conclusion from the assessment is that many different policy instruments can be used to further develop the waste management system in a more sustainable direction. In this section we discuss the scope for considering different types of policy combinations. The rationales for relying on policy mixes rather than isolated policy instruments are several. One argument, typically highlighted in environmental economics literature, is the difficulty in monitoring individual recycling behavior and thus the presence of illegal dumping. This calls, as argued above, for a combined output tax and recycling subsidy (equivalent to a deposit-refund system) [73]. Moreover, several policy instruments affect different parts of the waste management system and may address different types of market failures. Apart from environmental effects, policies are needed to address information problems, recycling design incentives etc. (e.g., [92]). Waste management policies must also take into account different political constraints and issues of public acceptance (apart from economic efficiency concerns and environmental effectiveness) (e.g., [102]).

Among the studied policy instruments, one in particular has the potential to significantly reduce the environmental impacts: "Required recycling of recyclable materials", provided that the recycled material can replace virgin material. In Sweden the bans of landfill disposal of organic and combustible materials has moved a lot of waste one step up in the waste hierarchy, from landfill to incineration with energy recovery. A requirement to recycle materials that can be recycled would move waste an additional step up in the hierarchy. This suggestion is also in line with the European Commission road map for a resource efficient Europe. The assessment presented here indicates that the costs are not prohibitory high compared to the environmental benefits. Before being implemented it will, however, need more assessments looking into costs and markets of recycled materials.

Moreover, many of the discussed policy instruments steer in a more sustainable direction although the impacts of several of these are rather limited. This suggests that many instruments need to be used in combination. It also suggests that they should be adjusted and some of them need to be further developed.

An effective strategy for decreasing the environmental impacts of waste management is in most cases to increase recycling. What policy instruments contribute most to an increase in the global recycling rates depends on whether or not there is a well-established, international recycling market. For materials where such a market exists, Swedish policy instrument should primarily focus on increasing the collection for recycling. The collected recyclables can be assumed to displace virgin materials in the international market where the two compete. This is because the marginal production of the material is likely to be based on virgin materials.

A policy instrument that focuses on increasing the use of recycled materials only in Sweden may be relatively ineffective in a situation where the established recycling markets are international. In such situations, an isolated support to the use of recycled materials in Sweden, may result in a lower recycling in other countries, and therefore not a globally increased recycling. It should therefore preferably be combined with policy instruments supporting the supply of recyclable materials, in order to increase global recycling [50].

A broad range of policy instruments are available to increase collection for recycling and thus the supply of recyclable materials. These include, for example, requirements to recycle recyclable materials, weight-based waste-collection fees, broader and tighter extended producer responsibility, requirements on curbside collection, deposit-refund systems, requirements on design for recycling, etc.

For materials where a recycling market does not exist, or when it is not well established, the reason is often a low demand for the recycled material. In such cases, the policy instrument could focus also on stimulating the demand for the recycled material, thus helping to establish the market. Examples of initiatives and policies that can be used to help establishing recycling markets include support for developing new recycling technologies (e.g., pilot and demonstration plants) and initiatives to decrease transaction costs. Green public procurement requirements can be used to demand a certain amount of recycled material in products and materials, which may be instrumental in developing a market.

A tax on virgin raw materials is another possible policy instrument that could support the establishment of recycling markets. It will not force recycling into existence, but it might stimulate technology and systems that make recycling feasible. Also in this case, supply-oriented policies would be a useful complement.

Reducing the amount of waste is top priority in waste policies and it can be done in several different ways. One may be to reduce the production volume and thus the waste from the production and after consumption. Another may be to change the production from more waste intensive products and services to less. A third is to change the production process to become less waste intensive. A fourth may be to make sure that products and materials are re-used or recycled before they become waste.

Increasing the costs of waste disposal could, in theory, reduce the amount of waste by all the mechanisms described above. Our general equilibrium model results indicate, however, very little to no changes in waste volumes resulting from changes in the marginal costs of waste disposal [24]. These results indicate that changes in the waste-management system may have little effect on the production, unless costs are raised to a much higher level than today. This outcome is reasonable, because even an increased cost of waste management constitutes just a small share of the overall cost of most businesses. Increased waste disposal costs could, however, still influence the waste treatment, for example by making recycling more economically attractive.

Another strategy to reduce the generated waste amounts may be to stimulate material efficiency by increasing the purchase cost of materials. This could be done by introducing raw material taxes which would make virgin raw material more expensive and/or by material taxes that hits both virgin and recycled materials. Another way of making materials more expensive is to tax environmental externalities from raw material production. For example, industrial wastes, including mining wastes, are largely exempt from the landfill tax in Sweden. If landfilled mining wastes would be taxed, this would increase the raw material costs in a similar way to the raw material tax analyzed here. The advantages and disadvantages would therefore be similar to those discussed above in relation to raw material tax.

As indicated by the discussion, a number of policy instruments can be used to support an increased recycling and decreased waste generation. The appropriate mix may depend on whether there are established recycling markets or not and, if not, what are the barriers for establishing such markets. If a recycling market does not exist today for a specific material, it might still be established in the future. In fact, one of the aims of some of the policy instruments could be to stimulate the establishment of such markets. A dynamic set of policy instruments could be designed as follows:

1. The material in question cannot be recycled. Technological development is supported to find recycling possibilities with environmental improvements. A tax on virgin raw materials is introduced to stimulate the development of technology and market for recycling.
2. Recycling of the material is possible but markets are not established. Further technological development is supported as well as tools to establish markets: for example, information systems, certifications, procurement requirements, waste brokers, and requirements for design for recycling.
3. A market has been established. A requirement to recycle is introduced together with other policy tools that support the supply of recyclable materials.

The policies evaluated above that focus on reducing waste amounts, have in general rather limited impacts. In order to radically decrease the production of waste, and transform norms and habits, more transforming policy instruments may therefore be needed, i.e., instruments that can cause social and cultural change and break current trends. Policy instrument that could start or enhance such change need to be perceived and experienced as meaningful for the recipients. This means that the instruments must be grounded in everyday life and 'mind'. They must correspond to something that, however costly, is fairly straightforward to do and understand, and it must give a real or abstract gain. If these criteria are met it means that the instruments should be intelligible, comprehensible and perceived as legitimate [103]. Even if an instrument is perceived radical, and therefore meets resistance, it does not mean that it is wrong to introduce it to begin with. In this case, the policy instrument could function as a 'stage for reflection' [104].

When introducing a radical measure, significant social consequences could in many cases be expected. There is always a risk for a backlash, with people finding ways not to co-operate or comply with the measures. But in addition to protest the affected public is also likely to be stimulated to come up with ideas and demand improved or new options for patterns of actions in everyday life. This means that parallel introduction of supplementary measures could seem reasonable in the public view. Such measures could also be used to adjust for perceived unfair distribution effects (of costs and inconveniences). However, somewhat paradoxically, negative publicity from features such as dramatic political processes or conflicting interests can lead to debate and media coverage that enhance public knowledge concerning how the instrument works. Such knowledge could actually pave the way for the understanding of a policy instrument and its implementation [105].

Today's well working, and publicly accepted, policy instruments seem to be grounded in a certain kind of reciprocity, 'the authorities arrange environmentally sound recovery so I pull my weight' [103], or more generally, people perceive they give something (i.e., pay) to receive something else [106]. From the ethnological point of view, policy instruments could be seen as a way of negotiating environmental problems or conflicts [107]. Policy instruments could actually be a rather good way to create large-scale cooperation. If we are to see policy instruments as a way of negotiating, then the negotiating parties are industry, authorities and people, with the internal conflict between the roles of the citizen and the more self-interested consumer or producer/employee [103]. The very notion of the environmental protection also plays a role (c.f. [47]).

When looking at current waste handling we find that sorting packaging to some extent is no longer being negotiated and therefore can be said to be agreed upon [43]. This indicates that authorities who introduce and maintain policy instruments have successfully mediated between citizen interest and consumer interest. Both the citizen and the consumer seem to benefit.

Correspondingly the conditions for waste minimization are still not agreed upon and must be settled in future negotiations. If we view a policy instrument as a mediator in a negotiation between different parts in a conflict, it is important that the mediator does not lose its status/power.

When we studied implementation of policy instruments such as recycling of packaging, we came to the conclusion that policy instruments and their contexts will always be in a state of flux. This means that they need to be maintained, adjusted and complemented at intervals. Governance of an environmental policy instrument must be active [103]. Solutions should to our minds be seen as associated with the negotiation of rules, restrictions and sanctions for the protection and management of natural resources [108,109]. We have found it important to start to qualitatively assess how the regulations function for people's (culture-characterized) ways of thinking and acting. The concept negotiation provides a perspective on the relationship between actors with the power to introduce regulations and the groups whose everyday lives are affected.

In the long term, it will not be possible to 'entice' people to conform to the regulations and the system if these do not produce the promised or expected outcome. Therefore we need to regularly determine whether people perceive that the regulations and systems fulfill their function [103]. This should be assessed both on the societal level and in nature. From a scientific point of view, improved theoretical frameworks, methods and interdisciplinary syntheses are needed to make better such assessments. From a societal point of view co-evolution of policies, technology and socio-cultural practices are needed to achieve waste minimization and sustainable waste management.

Many of the policy tools discussed above show interesting results in the evaluations. Some focus on preventing waste to arise in the first place, while others focus mainly on managing waste that has arisen. In order to significantly reduce waste amounts, more radical changes may be needed. That is why we have also discussed aspects of transforming policy instruments that break current trends. But more work is also needed before a detailed design of such tools can be suggested.

9.5 CONCLUSIONS

Through the assessments and lessons learned in the research program TOSUWAMA, we can establish that several policy instruments can be effective and lead towards a more sustainable waste management system. They also seem possible to implement. Particularly, we put forward the policy instruments "Information"; "Compulsory recycling of recyclable materials"; "Weight based waste fee in combination with information and developed recycling systems"; "Mandatory labeling of products containing hazardous chemicals", "Advertisements on request only and other waste minimization measures"; "Differentiated VAT and subsidies for some services". Compulsory recycling of recyclable materials is the policy instrument that has the largest potential of decreasing the environmental impacts. The effect of the other policy instrument as evaluated here could be limited and they need to be redesigned or used in combination in order to reduce environmental impacts more significantly. Furthermore, efforts are needed to take into account market and international aspects. In the more long term perspective, this set of policy instruments may need to be complemented with more transformational policy instruments that can significantly decrease the generation of waste.

REFERENCES

4. European Commission. Roadmap to a Resource Efficient Europe; COM (2011) 571 Final; European Commission: Brussels, Belgium, 2011.
5. Rockström, J.; Steffen, W.; Noone, K.; Persson, Å.; Chapin, III, F.S.; Lambin, E.F.; Lenton, T.M.; Scheffer, M.; Folke, C.; Schellnhuber, H.J.; et al. A safe operating space for humanity. Nature 2009, 461, 472–475.
6. Swedish EPA. Steg på vägen. Fördjupad utvärdering av miljömålen 2012(Step by Step. In Depth Evaluation of Environmental Objectives 2012); Report 6500; Swedish EPA: Stockholm, Sweden, 2012.
7. Ambell, C.; Björklund, A.; Ljunggren Söderman, M. Potential för ökad materialåtervinning av hushållsavfall och industriavfall (Potential for Increased Material Recycling of Household Waste and Industrial Waste); TRITA-INFRA-FMS 2010:4; KTH Samhällsplanering: Stockholm, Sweden, 2010.
8. Bernstad, A.; La Cour Jansen, J.; Aspegren, H. Life cycle assessment of a household solid waste source separation programme: A Swedish case study. Waste Manage. Res. 2011, 29, 1027–1042.

9. Björklund, A.; Finnveden, G. Life cycle assessment of a national policy proposal—The case of a proposed waste incineration tax. Waste Manage. 2007, 27, 1046–1058.
10. European Environment Agency (EEA). The Road from Landfilling to Recycling: Common Destination, Different Routes; European Environment Agency: Copenhagen, Denmark, 2007. ISBN 978-92-9167-930-0.
11. Pires, A.; Martinho, G.; Chang, N-B. Solid waste management in European countries: A review of systems analysis techniques. J. Environ. Manage. 2011, 92, 1033–1050.
12. Gentil, E.; Clavreul, J.; Christensen, T.H. Global warming factor of municipal solid waste management in Europe. Waste Manage. Res. 2009, 27, 850–860.
13. EU. Directive 2008/98/EC of the European Parliament and of the Council of 19 November 2008 on Waste and Repealing Certain Directives; European Commission: Brussels, Belgium, 2008.
14. Swedish Government. Svenska miljömål—preciseringar av miljökvalitetsmålen och en första uppsättning etappmål (SwedishEnvironmental Objectives—Clarifications of the Envrionmental objectIves and a First Set of Milestones); Ds 2012:23; Regeringskansliet: Stockholm, Sweden, 2012.
15. Swedish Waste Management. Svensk avfallshantering 2010 (Swedish waste management 2010); Swedish Waste Management: Malmö, Sweden, 2011.
16. Ljunggren Söderman, M. Assessment of Policy Instruments for Waste Prevention. In Presentation at the ISWA Beacon Conference on Waste Prevention and Recycling, Vienna, Austria, 2011; ISWA: Vienna, Austria, 2011.
17. Swedish EPA. Avfall i Sverige 2010 (Waste in Sweden 2010); Report 6520; Swedish EPA: Stockholm, Sweden, 2012.
18. Swedish EPA. Från avfallshantering till resurshushållning. Sveriges avfallsplan 2012–2017 (From Waste Management to Resource Management. Waste plan of Sweden 2012–2017); Swedish EPA: Stockholm, Sweden, 2012.
19. Von Borgstede, C.; Andersson, K. Environmental information—Explanatory factors for information behavior. Sustainability 2010, 2, 2785–2798.
20. Henriksson, G.; Åkesson, L.; Ewert, S. Uncertainty regarding waste handling in everyday life. Sustainability 2010, 2, 2799–2813. [Google Scholar] [CrossRef]
21. Swedish District Heating Association. 2012. Available online: http://www.svenskfjarrvarme.se/Statistik--Pris/Fjarrvarme/Energitillforsel/ (accessed on 14 November 2012).
22. Towards Sustainable Waste Management. 2013. Available online: http://www.hallbaravfallshantering.se (accessed on 21 February 2013).
23. Bisaillon, M.; Finnveden, G.; Noring, M.; Stenmarck, Å.; Sundberg, J.; Sundqvist, J.-O.; Tyskeng, S. Nya styrmedel inom avfallsområdet? (New Policy Measures for Waste Management?); Miljöstrategisk analys—fms, ISSN 1652-5442, TRITA-INFRA-FMS 2009:7; KTH Royal Institute of Technology: Stockholm, Sweden, 2009.
24. Finnveden, G.; Bisaillon, M.; Noring, M.; Stenmarck, Å.; Sundberg, J.; Sundqvist, J.-O.; Tyskeng, S. Developing and evaluating new policy instruments for sustainable waste management. Int. J. Environ. Sust. Develop. 2012, 11, 19–31.

25. Dreborg, K.-H.; Tyskeng, S. Framtida förutsättningar för en hållbar avfallshantering—Övergripande omvärldsscenarier samt referensscenario(Future conditions for a sustainable waste management—General environment scenarios and reference scenario); Miljöstrategisk analys—fms, TRITA-INFRA-FMS 2008:6; KTH Royal Institute of Technology: Stockholm, Sweden, 2008.
26. Sjöström, M.; Östblom, G. Future Waste Scenarios for Sweden Based on a CGE-model; Working Paper no. 109; National Institute of Economic Research: Stockholm, Sweden, 2009.
27. Östblom, G.; Ljunggren Söderman, M.; Sjöström, M. Analysing Future Waste Generation—Soft Linking a Model for Waste Management with a CGE-model for Sweden; Working paper no. 118; National Institute of Economic Research: Stockholm, Sweden, 2010.
28. Ljunggren Söderman, M.; Björklund, A.; Eriksson, O.; Forsfält, T.; Stenmarck, Å.; Sundqvist, J.-O. Policy Instruments for a More Sustainable Waste Management; Presentation at LCM 2011; Berlin, Germany, 2011. Available online: http://www.lcm2011.org (accessed on 21 February 2013).
29. Östblom, G.; Berg, C. The EMEC model: Version 2.0; Working Paper no. 96; National Institute of Economic Research: Stockholm, Sweden, 2006.
30. Forsfält, T. Samhällsekonomiska effekter av två styrmedel för minskade avfallsmängder (Socioeconomic impacts of two policy measures towards reduced waste amounts); Specialstudie nr 26; National Institute of Economic Research: Stockholm, Sweden, 2011.
31. Ljunggren, M. Modelling national solid waste management. Waste Manage. Res. 2000, 18, 525–537.
32. Ljunggren Söderman, M. Ekonomisk analys av nya styrmedel för hanteringen av svenskt avfall(Economic Analysis of New Policy Measures for Management of Swedish Waste); Report B 2021; IVL Swedish Environmental Research Institute: Stockholm, Sweden, 2011.
33. Arushanyan, Y.; Björklund, A.; Eriksson, O.; Finnveden, G.; Ljunggren-Söderman, M.; Sundqvist, J.-O.; Stenmarck, Å. Environmental Assessment of Waste Policy Instruments in Sweden; 2013. in progress.
34. ISO. ISO 14040 International Standard. Environmental Management—Life Cycle Assessment —Principles and Framework; International Organisation for Standardization: Geneva, Switzerland, 2006.
35. Clift, R.; Doig, A.; Finnveden, G. The Application of Life Cycle Assessment to Integrated Solid Waste management, Part I—Methodology; Process Safety and Environmental Protection 2000, 78, 279–287.
36. Finnveden, G.; Hauschild, M.; Ekvall, T.; Guinée, J.; Heijungs, R.; Hellweg, S.; Koehler, A.; Pennington, D.; Suh, S. Recent developments in Life Cycle Assessment. J. Environ. Manage. 2009, 91, 1–21.
37. Pré, Simapro 7 Professional Version 7.3.2. Pre Consultants, Amersfoort, The Netherlands, 2011.
38. Goedkoop, M.; Heijungs, R.; Huijbregts, M.; De Scryver, A.; Sruijs, J.; van Zelm, R. ReCiPe 2008, Report 1: Characterization; Ministry of VROM: The Hague, The Netherlands, 2009.

39. Frischknecht, R.; Jungbluth, N.; Althaus, H.; Doka, G.; Dones, R.; Hellweg, S.; Hischier, R. Implementation of Life Cycle Impact Assessment Methods; Swiss Center for Life Cycle Inventories: Duebendorf, Switzerland, 2007.
40. Bösch, M.; Hellweg, S.; Huijbregts, M.; Frischknecht, R. Applying cumulative exergy demand (CExD) indicators to the ecoinvent database. Int. J. Life Cycle Ass. 2006, 12, 181–190.
41. Ekvall, T. Key methodological issues for life cycle inventory analysis of paper recycling. J. Clean. Prod. 1999, 7, 281–294.
42. Wene, C.-O. Exploring and Mapping: A Comparison of the IEA-MARKAL and CEC-EFOM Technical Energy System Models and the ANL Electric Utility Model; Technical report BNL-52224; Brookhaven National Lab.: Upton, NY, USA, 1989.
43. Sahlin, J.; Ekvall, T.; Bisaillon, M.; Sundberg, J. Introduction of a waste incineration tax: Effects on the Swedish waste flows. Resour. Conserv. Recy. 2007, 51, 827–846.
44. Blomberg, J.; Söderholm, P. The economics of secondary aluminium supply: An econometric analysis based on European data. Resour. Conserv. Recy. 2009, 53, 455–463.
45. Ekvall, T.; Sahlin, J.; Sundberg, J. Effects of Policy Instruments on Waste Intensities; Report B1939; IVL Swedish Environmental Research Institute: Stockholm, Sweden, 2010.
46. Henriksson, G.; Åkesson, L.; Ewert, S. Uncertainty regarding waste handling in everyday life. Sustainability 2010, 2, 2799–2813.
47. Andersson, M.; von Borgstede, C.; Eriksson, O.; Guath, M.; Henriksson, G.; Sundqvist, J.-O.; Åkesson, L. Hållbar avfallshantering—utvärdering av styrmedel från psykologiskt och etnologiskt perspektiv (Sustainable Waste Management—Evaluation of Policy Measures from an Psychological and Ethnologivcal Perspective); TRITA-INFRA-FMS: 2011:5; KTH Royal Institute of Technology: Stockholm, Sweden, 2011.
48. Andersson, M.; Eriksson, O.; von Borgstede, C. The effects of environmental management systems on source separation in the work and home settings. Sustainability 2012, 4, 1292–1308.
49. Geertz, C. The Interpretation of Cultures; Basic Books: New York, NY, USA, 1973.
50. Gram-Hanssen, K. Practice Theory and the Green Energy Consumer. In Presentation at ESA Conference 3–6 September 2007 in Glasgow, Research Network on the Sociology of Consumption, 2007; ESA: Paris, France, 2007.
51. Ekvall, T.; Sundqvist, J.-O.; Hemström, K.; Jensen, C. Stakeholder Analysis of Incineration Tax, Raw Material Tax, and Weight-based Waste Fee; Draft report; IVL Swedish Environmental Research Institute: Stockholm, Sweden, 2011.
52. Söderholm, P.; Ekvall, T. Material Markets in the Presence of Secondary and Primary Production: Important Interactions and Policy Impacts, 2012. Draft version.
53. Söderholm, P. Taxing Virgin Natural Resources: Lessons from Aggregates Taxation in Europe. Resour. Conserv. Recy. 2011, 55, 911–922.
54. Olofsson, M.; Sahlin, J.; Ekvall, T.; Sundberg, J. Driving forces for import of waste for energy recovery in Sweden. Waste Manage. Res. 2005, 23, 3–12.

55. Bergek, A.; Jacobsson, S. Are tradable green certificates a cost-efficient policy driving technical change or a rent-generating machine? Lessons from Sweden 2003–2008. Energ. Policy 2010, 38, 1255–1271.
56. Swedish Waste Management. National Survey of Analyses of Composition of Household's Solid Waste; Report U 2011:4; Swedish Waste Management: Malmö, Sweden, 2011.
57. Massachusetts Department of Environment Protection. Massachusetts 2010–2020. Solid Waste Master Plan. Pathway to Zero Waste; Massachusetts Department of Environment Protection: Boston, MA, USA, 2010.
58. Gerlat, A. Mandatory Organics Recycling to Become Law in Vermont. 2012. Available online: http://waste360.com/state-and-local/mandatory-organics-recycling-become-law-vermont (accessed on 14 November 2012).
59. Metro Vancouver. Banned and Prohibited Materials. 2012. Available online: http://www.metrovancouver.org/services/solidwaste/disposal/Pages/bannedmaterials.aspx (accessed on 14 November 2012).
60. Ahlroth, S.; Finnveden, G. Ecovalue08—A new valuation set for environmental systems analysis tools. J. Clean. Prod. 2011, 19, 1994–2003.
61. Arushanyan, Y.; Finnveden, G. Ban on incineration, weighting gf.xlsx; Spreadsheet Data, 20120222, Stockholm, Sweden, 2012. Unpublished.
62. Nilsson, H. Förpacknings- och tidningsinsamlingen AB, Stockholm, Sweden. Personal communication, 13 December 2011.
63. Profu, Tillgång och efterfrågan på behandlingskapacitet för brännbart och övrigt organiskt avfall. Underlag till Sveriges nationella avfallsplan 2011—Del 1. (Supply and Demand for Treatment Capacity of Combustible and Other Organic Waste. Data for Sweden's National Waste Plan 2011—Part 1), Profu, Mölndal, Sweden, 1 April 2011.
64. Fischer, J. Massachusetts Department of Environment Protection, Boston, MA, USA. Personal Communication, 22 February 2012.
65. O'Doherty, R.; Bailey, I.; Colins, A. Regulatory failure via market evolution: The case of UK packaging recycling. Environ. Plann. C Govern. Pol. 2003, 21, 579–595.
66. Matsueda, N.; Nagase, Y. An economic analysis of the packaging waste recovery note system in the UK. Resour. Energy Econ. 2012, 34, 660–679.
67. Edjemo, T.; Söderholm, P. Steel Scrap Markets in Europe and the USA. Miner. Energ. 2008, 23, 57–73.
68. Weitzman, M. Prices vs. Quantities. Rev. Econ. Stud. 1974, 41, 447–491.
69. Björklund, A.; Finnveden, G. Life cycle assessment of a national policy proposal—The case of a proposed waste incineration tax. Waste Manage. 2007, 27, 1046–1058.
70. SOU, Skatt i retur (Tax in return); SOU 2009:12, Fritzes, Stockholm, Sweden, 2009.
71. Dahlén, L.; Lagerkvist, A. Pay as you throw: Strengths and weaknesses of weight-based billing in household waste collection systems. Waste Manage. 2010, 30, 23–31.
72. Hage, O.; Sandberg, K.; Söderholm, P.; Berglund, C. Household Plastic Waste Collection in Swedish Municipalities: A Spatial-Econometric Approach. In

Proceedings of The 16th Annual Conference of the European Association of Environmental and Resource Economists (EAERE), Gothenburg, Sweden, 25–28 June 2008; EAERE: Venice, Italy, 2008.
73. Hedman, B.; Näslund, M.; Nilsson, C.; Marklund, S. Emissions of polychlorinated Dibenzodioxins and Dibenzofurans and Polychlorinated Biphenyls from uncontrolled burning of garden and domestic waste (Backyard Burning). Environ. Sci. Technol. 2005, 39, 8790–8796.
74. Schmidt, L.; Sjöström, J.; Palm, D.; Ekvall, T. Viktbaserad avfallstaxa—Vart tar avfallet vägen? (Weight-Based Waste Tariff—Where does the Waste Go?); Report B 2054; IVL Swedish Environmental Research Institute: Stockholm, Sweden, 2012. [Google Scholar]
75. Sterner, T.; Bartelings, H. Household waste management in a Swedish Municipality: Determinants of waste disposal, Recycling and composting. Environ. Resour. Econ. 1999, 13, 473–491. [Google Scholar] [CrossRef]
76. Fullerton, D.; Kinnaman, T.C. Garbage, Recycling and Illicit Burning or Dumping. J. Environ. Econ. Manag. 1995, 29, 78–91. [Google Scholar] [CrossRef]
77. Walls, M.; Palmer, K. Upstream pollution, Downstream waste disposal, And the design of comprehensive environmental policies. J. Environ. Econ. Manag. 2001, 41, 94–108.
78. Environmental Policy and Household Behavior: Sustainability and Everyday Life; Söderholm, P., Ed.; Earthscan: London, UK, 2010.
79. Thøgersen, J. Monetary incentives and environmental concern. Effects of a differentiated garbage fee. J. Consum. Policy 1994, 17, 407–443.
80. Kinnaman, T.C. Policy Watch: Examining the Justification for Residential Recycling. J. Econ. Perspec. 2006, 20, 219–232.
81. Swedish EPA. System för insamling av hushållsavfall i materialströmmar(System for collection of household waste in material streams); Report 5942; Swedish EPA: Stockholm, Sweden, 2009.
82. Palmer, K.; Sigman, H.A.; Walls, M. The Cost of Reducing Municipal Solid Waste. J. Environ. Econ. Manag. 1997, 33, 128–150.
83. Ljunggren Söderman, M.; Björklund, A. Konsumtion, produktion och framtida avfall—effekter på miljö och ekonomi (Consumption, Production and Future Waste-Effects on the Environment and Economy). Presentation at the conference Avfall i nytt fokus, Borås, Sweden, 2010. Available online: http://www.hallbaravfallshantering.se/ (accessed on 21 February 2013).
84. Brekke, K.A.; Kipperberg, G.; Nyborg, K. Social interaction in responsibility ascription: The case of household recycling. Land Econ. 2010, 86, 766–784.
85. Bruvoll, A.; Nyborg, K. The cold shiver of not giving enough: On the social cost of recycling campaigns. Land Econ. 2004, 80, 539–549.
86. Cela, E.; Kaneko, S. Determining the effectiveness of the Danish packaging tax policy: The case of paper and paperboard and packaging imports. Resour. Conserv. Recy. 2011, 55, 836–841.
87. Rouw, M.; Worrell, E. Evaluating the impacts of packaging policy in The Netherlands. Resour. Conserv. Recy. 2011, 55, 483–492.

88. Monomaivibool, V.; Vassanadumrongdee, S. Extended Producer Responsibility in Thailand. Prospects for Policies on Waste Electrical and Electronic Equipment. J. Ind. Ecol. 2011, 15, 185–205.
89. Mayers, K.; Peagam, R.; France, C.; Basson, L.; Clift, R. Redesigning the Camel: The European WEEE directive. J. Ind. Ecol. 2011, 15, 4–8.
90. Lindhqvist, T. Policies for waste batteries: Learning from experience. J. Ind. Ecol. 2010, 14, 537–540.
91. Cleary, J. Life cycle assessments of wine and spirit packaging at the product and municipal scale: A Toronto, Canada case study. J. Clean. Prod. 2013. in press.
92. Gentil, E.; Gallo, D.; Christensen, T.H. Environmental evaluation of municipal waste prevention. Waste Manage. 2011, 31, 2371–2379.
93. Ljunggren Söderman, M.; Davidsson, H.; Jensen, C.; Palm, D.; Stenmarck, Å. Goda exempel på förebyggande av avfall från kommuner (Good Examples of Waste Prevention in Municipalities); Report U 2011:5; Swedish Waste Management: Malmö, Sweden, 2011.
94. Salhofer, S.; Obersteiner, G.; Schneider, F.; Lebersorger, S. Potential for the prevention of municipal solid waste. Waste Manage. 2008, 28, 245–259.
95. Nicolli, F.; Johnstone, N.; Söderholm, P. Resolving Failures in Recycling Markets: The Role of Technological Innovation. Environ. Econ. Policy Stud. 2012, 14, 261–288.
96. Watkins, G.; Husgafvel, R.; Pajunen, N.; Dahl, O.; Heiskanen, K. Overcoming institutional barriers in the development of novel process industry residue based symbiosis product—Case study at the EU level. Miner. Eng. 2013, 41, 31–40.
97. Fullerton, D.; Wu, W. Policies for Green Design. J. Environ. Econ. Manag. 1998, 25, 242–256.
98. Calcott, P.; Walls, M. Waste, Recycling, and 'Design for Environment': Roles for Markets and Policy Instruments. Resour. Energ. Econ. 2005, 27, 283–305.
99. Johansson, J.G.; Björklund, A.E. Reducing lifecycle environmental impact of waste electrical and electronic equipment recycling. J. Ind. Ecol. 2010, 14, 258–269.
100. Söderholm, P. Economic Instruments in Chemicals Policy: Past Experiences and Prospects for Future Use; TemaNord 2009:565; Nordic Council of Ministers: Copenhagen, Denmark, 2009.
101. Tyskeng, S.; Finnveden, G. Comparing energy use and environmental impacts of recycling and incineration. J. Environ. Eng. 2010, 136, 744–748.
102. Nakatani, J.; Fujii, M.; Moriguchi, Y.; Hirao, M. Life-cycle assessment of domestic and transboundary recycling of post-consumer PET bottles. Int. J. Life Cycle Ass. 2010, 15, 590–597.
103. Profu, Evaluating Waste Incineration as Treatment and Energy Recovery Method from an Environmental Point of View; Report on behalf of CEWEP (Confederation of European Waste-to-Energy Plants), Final Version, 13 May 2004, Profu, Mölndal, Stockholm, 2004.
104. Eriksson, O.; Finnveden, G. Plastic waste as a fuel—CO2-neutral or not? Energ. Environ. Sci. 2009, 2, 907–914.
105. Damon, M.; Sterner, T. Policy Instruments for sustainable development at Rio + 20. J. Environ. Develop. 2012, 21, 143–151.

106. Henriksson, G.; Börjesson Rivera, M.; Åkesson, L. Environmental policy instruments seen as negotiations. In Negotiating Environmental Conflicts: Local communities, global policies (published within the series Kulturanthropologie Notizen, vol. 81); Goethe Universität: Frankfurt am Main, Germany, 2012.
107. Hajer, M.A. A Frame in the Fields: Policymaking and the Reinvention of Politics. In Deliberative Policy Analysis: Understanding Governance in the Network Society; Hajer, M.A., Wagenaar, H., Eds.; Cambridge University Press: Cambridge, UK, 2003; pp. 88–110.
108. Henriksson, G. What did the Trial Mean for Stockholmers? In Congestion Taxes in City Traffic: Lessons Learnt from the Stockholm Trial; Gullberg, A., Isaksson, K., Eds.; Nordic Academic Press: Lund, Sweden, 2009; pp. 235–294.
109. Shove, E.; Walker, G. Governing transistions in the sustainability of everyday life. Res. Policy 2010, 39, 471–476.
110. Gulliver, P.H. Anthropological contributions to the study of negotiations. Negotiation J. 1988, 4, 247–255.
111. Worster, D. Nature's Economy. A History of Ecological Ideas, 2nd ed.; Cambridge University Press: Cambridge, UK, 1996.
112. Ostrom, E. Coping with tragedies of the commons. Annu. Rev. Polit. Sci. 1999, 2, 493–535.

CHAPTER 10

Framework for Low Carbon Precinct Design from a Zero Waste Approach

QUEENA K QIAN, STEFFEN LEHMANN, ATIQ UZ ZAMAN, AND JOHN DEVLIN

10.1 INTRODUCTION

The development of science and technology as well as global levels of economic activity causes a dramatic increase in the production of urban solid waste [1]. The generation of waste over time has become a serious environmental problem for the world, and been affecting the balance of natural resources [2]. Solid Waste Management (SWM) has become crucial for protecting the environment and the human wellbeing. Various national and international initiatives for SWM are in place, which takes considerations of environmental, administrative, regulatory, scientific, market, technology, institutional and socioeconomic factors [3].

The sustainable SWM is becoming essential at all phases of the waste chain from production, waste generation, collection, transportation, treatment, recycling till disposal. 'Zero waste' is, therefore, becoming a popular concept. It is a closed-loop concept aiming of optimum recycling or resource recovery, as well as elimination of unnecessary waste in the first place [4,5]. With a whole system approach, it seeks for an end–of-pipe solution for waste diversion along the materials flow through society. It encourages waste elimination through recycling and resource recovery, with a guiding design philosophy to reduce waste at source and at all points down the supply chain [6]. 'Zero waste' commitments have been made across the world, including US, Europe, Australia, New Zealand, etc. [7], and becomes trendy for the rest of the world.

A sustainable SWM approach is systematic, flexible and long term visionary. A sustainable society requires sophisticated ways to manage solid waste. A systems approach that reveals the relationships and explains its interactions among the parties in the system contributes to greater sustainable practice [8]. Based on reviewing and comparing different researchers' work on the waste management, our research aims to propose a research framework of zero-waste management and strategies for low carbon residential precincts. This approach selected needs to accommodate the fact that zero waste management can be achieved by identifying the leverage points during the entire zero waste chain and altering or redesigning the processes accordingly. Kytzia and Nathani believe a "combination between analyses of economic/ physical structures on the one hand and economic behaviour on the other hand is most promising" to achieve the zero waste concept [9]. The methodological framework presented will contributes to the understanding of the overall process of the zero waste management by combining system characteristics as well as the cost/ benefit impact with the attitudes and requirements of a specific stakeholder group (i.e., the city planner, government, and/or households). This paper highlights the dynamic interrelationships of the sustainable SWM practices, supplemented with the cost/benefit factors into the SD process. The system-oriented research framework serves the decisionmakers to draw the forward-looking and preventative insights and reach a scientific understanding of the carbon and cost consequences relating to various sustainable SWM scenarios.

10.2 LITERATURE REVIEW

10.2.1 SIMILAR RESEARCH IN SWM AND SYSTEM APPROACH

Research interests in addressing waste management issues have resulted in a large amount of publications during the last decade. Ossenbruggen and

Ossenbruggen apply SWAP programs - a linear programming algorithm, to aid in the strategic plan and decisionmakings of SWM, and weigh the cost associated and the benefits from various waste recovery alternatives [10]. Chung and Poon has applied multiple criteria analysis (MCA) in SWM to find out the preferred waste management options. The merit of MCA is more objective and transparent and it accommodates quantitative and qualitative data [11]. Bovea et al. apply the Life Cycle Assessment technique to obtain parameters that quantifies the environmental impact of waste transportation and operating a transfer station in municipal SWM systems [12]. Beigl et al. review the modelling approaches for SWM and propose an implementation guideline with a compromise between information gain and cost-efficient model development [13]. Lu and Yuan develop a framework to understand the C and D WM research as archived in selected journals, and give useful references attempting the research of C and D WM research [14]. Chang and Davila simulate the predetermined scenarios with a minimax regret optimization, to achieve improved SWM strategies from different environmental, economic, legal, and social conditions [15].

Planning sustainable SWM has to address several interdependent issues including public health, the environmental impact, the treatment potential, the landfill capacity, and present and future economic and social costs, and financial expenditures, etc. It, therefore, becomes increasingly necessary to understand the dynamic nature of their interactions, and the complex, and multi-faceted system. How to combine all the correlated factors into the consideration when making the optimal sustainable SWM strategy among the alternatives? Pries et al. believes that system approach enriches the analytical framework of SWM, specially designed to understand the dynamics and intercalations among the factors, and develop better SWM strategies for both the SWM industry and the government [16]. It plays an important role to simulate and assess the integrated SWM systems, and inform the stakeholders with insightful strategies and rational decision-makings.

10.2.2 SYSTEM DYNAMICS (SD) APPROACH IN MSW MANAGEMENT

SD is a well-established methodology that provides a theoretical framework and concepts for modelling complex social, economic and managerial systems [17]. It deals with the interrelationships and complex of the system, where the dynamic behaviour can be reflected and simulated by the feedback loops based on the control theory [18–21]. The SD approach is widely applied in the areas of environmental sustainability and regional sustainable development issues [22-26] environmental management and environmental systems [27,28], and waste management [29-31].

Thirumuthy applied SD approach to evaluate the investments required for various environmental services in Madras city [32]. Mashayekhi explored a dynamic analysis for analysing the transition from the landfill method of disposal to other forms of disposal for the city of New York [33]. Sudhir et al. proposed a system dynamics model to capture the dynamic nature of interactions among the various components in SWM for developing countries [28]. Karavezyris et al. studied the quantitative impact of different variables, such as voluntary recycling participation and regulation, on SWM [34]. Ulli- Beer presented a SD model for understanding local recycling systems [35]. Dyson and Chang applied a SD approach to predict solid waste generation in a fast growing urban area [36]. Duran et al. developed a model to assess the economic viability of creating markets for recycled construction and demolition (CandD) waste in scenarios using different economic instruments [37]. Rehan et al. proposes SD approach to develop a causal loop diagram for water and wastewater network management, as a complex system with multiple interconnections and feedback loops [38]. It demonstrates the significance of feedback loops for financial management of the complexity of the system by incorporating all feedback loops. Yuan et al. proposes a SD model to serve as a decision support tool for waste projection, and as a platform for simulating effects of various waste reduction management strategies [39].

A sustainable SWM system incorporates feedback loops, focusing on processes, embodies adaptability and diverts wastes from disposal [8]. Due to the SWM hierarchy, the challenges lie in how to diversify the waste reduction options, increase the reliability of infrastructure systems, and leverage the redistribution of waste streams among production, transportation, compost, recycling, and other facilities. It depends on factors such as technology and infrastructure, socioeconomic and institutional, social- environment, culture, as well as market considerations [40,41]. Transitioning to a sustainable SWM system requires identification and application of leverage points that stimulate positive change [8]. SD approach recognizes the sustainable SWM process and accommodates the zero waste achievement by identifying the leverage points during the entire zero waste chain and altering or redesigning the processes accordingly.

10.2.3 COST-BENEFIT ANALYSIS (CBA) SUPPLEMENTED IN SD

Economic instruments to minimize waste play a crucial role in encouraging environmentally-friendly SWM practices [30]. The rising pressure in terms of cost efficiency on public services and facilities pushes governments to share those services with the industry. The partnership among the government, business and individual household collectively contributes to the overall effectiveness and efficiency of the MSW management. Therefore, CBA of waste management are essential to provide the evidence that will motivate stakeholders throughout waste management chains, and SD modelling helps examine the relationships between waste management activities in a holistic view. Yuan et al. analysed the CBA of the dynamics and interrelationships of C and D waste management practices using a SD approach [30]. Farel et al. proposes a SD approach to simulate the net economic balance of the recycling network under different future scenarios [42]. Therefore, a good balance between the cost and benefits is an important factor to select among the scenarios for the use of different stakeholders.

CBA has been widely acknowledged as a tool for policy and project analysis throughout the world. It helps the policy-makers as well as stakeholders to justify their decisions in a more systematic, rigorous and unambiguous way Gramlich [43]. It allows us to identify and assess positive and negative economic and physical effects independently. Particularly, it supports the simulation and optimization models for system analysis. Well-defined CBA parameters may translate environmental aspects into economic terms.

10.3 METHODOLOGICAL FRAMEWORK

10.3.1 THE LIFE CYCLE OF ZERO-WASTE IN DIAGRAM (FIGURE 1)

The waste management chain consists of a series of potential lifecycle stages. Products pass through the manufacturing, production, and consumption stages before entering into the waste management system where various processes can minimize the impact of the waste. It is not a collection of independent waste management activities but rather a system of interdependent activities. It is to put together all the potential acts to reduce the original waste to be generated, until the waste can finally be disposed of with minimum costs involved. Some typical factors affecting waste management activities are based on an extensive literature review as well as works previously carried out at the zero waste research centre [4].

10.3.2 SYSTEM DYNAMICS (SD) APPROACH-CAUSAL LOOP SD DIAGRAM FOR A ZERO-WASTE PRECINCT

This causal loop diagram is designed based on the interaction of different components in a zero waste management system. It identifies the carbon emission loops and external factors as well as interrelationships in SWM from waste generation (manufacture, household consumption, separation) to collection, treatment, recycling and disposal. This study frames the scope of the zero-waste loop and establishes the initial step of a SD study framework. It is used to identify and explain the causal relationships between acts and stages of waste management in a residential precinct. Based on the above extensive literature reviews on SD and conceptual model of the zero-waste management chain in low carbon residential precinct (Figure 1), the casual loop diagram of zero-waste precincts can be developed (Figure 2).

10.3.3 COST BENEFIT ANALYSIS (CBA) APPROACH

CBA is an analytical procedure to evaluate the desirability of a program or project by weighing the resulting benefits against the corresponding costs in order to see whether the benefits outweigh the costs [44-46]. In this research, CBA consider sand reflects the tangible factors (such as environmental costs) as well as invisible benefits, from the improved SWM scenarios. It attempts to evaluate effects on users (policy-makers as well as stakeholders), external effects, quantify values and social benefits. In this research, the current value of a collective SWM scenario is considered on a net present value (NPV) basis. NPV reflects a stream of current and future benefits and costs, and results in a value in today's dollars that represents the present value of an investment's future financial benefits minus any initial investment. Typically, financial benefits for individual elements are calculated on a present value basis and then combined in the conclusion with net costs to arrive to NPV, as the function below:

$$NPV = \sum_{i=1}^{n} \frac{values_1}{(1 + rate^i)} \tag{1}$$

If positive, the investment should be made, otherwise not. The value of the NPV of the proposed scenario gives a foundation for comparing alternative options. A bigger NPV indicates a better option.

Of particular interest to both waste industry stakeholders and policymakers, this research looks into both private cost-benefit ratios (to establish how much benefit can be derived from every dollar is spent for improving waste management) and social cost benefit ratios (as the amount of return is perceived to have a direct impact on the degree of success of the waste management regulation or incentive scheme). The amount of investment return of implementing a particular measure is evaluated using private benefit cost ratios. The mathematical function of adopting incentives can be obtained as following:

$$privatecost - benefiltratio_i^1 = \frac{\Delta B_i^{Econ}}{\Delta C_i^{Econ}} \tag{2}$$

Where ΔB_i^{Econ} and ΔC_i^{Econ} are the additional benefit derived, and additional life-cycle cost required for the private industry stakeholders and/or household by implementing the proposed waste management project or incentives, respectively. The private benefit–cost ratio can help identify which options are financially beneficial. The ratio can give an indication for selecting a particular measure if the ratio of the present value of benefits to the present value of costs is greater than 1.0.

$$socialcost - benefitratio_i^1 = \frac{\Delta B_i^{Soc}}{\Delta C_i^{Soc}} \quad (3)$$

Where ΔB_i^{Soc} is the net benefit in monetary value derived for society by implementing the proposed improvement measure; ΔC_i^{Soc} , which is the additional life-cycle cost required by implementing the proposed incentives. The social benefit–cost ratio can assist government officials or policy-makers in judging the environmental viability of the measure under consideration.

10.4 DISCUSSIONS AND FINDINGS

Table 1 lists the selection of SD waste management parameters base on the zero-waste casual loop diagrams (Figures 1 and 2), to be considered in CBA. All the parameters selected are based on the literature review and methodological framework done in the earlier session. They are assigned in the categories of Process, Technology and Infrastructure, Socio-economic, Institutional, and Socio-environment. All the parameters are assigned to be either under the cost (-) or benefit (+) or both, depending on whether they are mainly involving the costs or benefits, or both to the stakeholders along the SWM process. Very often, the parameters selected involve both costs and benefits depending on the different stakeholders.

The benefits to get CBA supplemented to the SD approach, in that it quantifies the potential returns and expenses of a program or project and balances the pros and cons to arrive at a decision. This line of research has important implications both for assessing the cost of correcting market failures – such as environmental externalities – as well as clarifying the role of policies that are oriented to correct behavioural failures and market barriers. The benefits and costs are usually quantified in real monetary terms, i.e. converted to the present value, to enable assessment of different the benefits and the costs over time. The major CBA indicators include present value, net present value and benefit cost ratio, which are applied in this study [47]. One of the aspects evaluated by CBA in this study is the hidden costs and benefits during the waste management loop, which has not been widely considered.

10.4.1 FUTURE STUDIES: SCENARIOS SETTING AND DATA COLLECTIONS—CASE OF BOWDEN VILLAGE, SA, AUSTRALIA

The framework proposed is to be tested in future study using reallife data, in three different residential density scenarios – mixed use high rise (10-storey), low rise (3 to 8-storey) and town house scale (3-storey). The data will be derived from Bowden Urban Village, a new residential precinct in Adelaide, the capital city of South Australia. Upon successful testing of the SD model using the Bowden village in future case study, it will be adjustable to a wider application to assist decision-making in different precincts. Most of the research on MSW management is examining systems at the city or national level. There are many benefits to this study's use of the precinct level: precincts provide more flexibility and precision, it also allows the model to be closer to reality as the dynamics and external factors at the city and/or national level are too complicated to be modelled.

The model described above is a theoretical framework for examining MSW loop and its management system at precinct level. For the future study case of Bowden urban village, three scenarios are set for different options and desire of the development plan- mixed use high rise (10-story), low rise (3 to 8-story) and town house scale (3-story). Using the SD simulation program, e.g. i think, it will examine the potential ways to reduce municipal solid waste (MSW) throughout generation, collection till disposal and concludes with overall carbon consequences and reduction options.

Following the general research framework developed in this paper, it is to collect and evaluate a broad spectrum of costs and benefits with the available data and the data from the survey, and to develop reasonable net present value estimation for comparison of each decision-making scenario. System dynamics approach is applied to capture and frame the scope of the waste life cycle in order to forecast the costs and benefits of waste management options from beginning to end. The overarching purpose is to answer the question: "does it make financial and economic sense for a given stakeholder, with particular incentives, to implement these waste management options?"; which serves the ultimate aim of this study: to balance the financial and economic interests of the private sector and those of the whole society (and environment) with minimum environmental impact and optimum social benefits, on behalf of the urban policy-maker.

10.5 CONCLUSIONS

Based on an extensive literature review on waste management of system approach, this paper proposes a holistic methodological framework for designing a SWM system for a zero-waste low carbon residential precinct. This study attempts to employ both a SD approach, incorporated with a cost-benefit analysis to simulate the changes in various MSW management scenarios for different lowcarbon precincts. The causal loop diagram made it easier to understand and identify the critical activities throughout the waste management chain, the essential stakeholders and also external factors such as financial incentives. The methodological framework considers a list of parameters under the categories of socio-economic, institutional, socio-environmental, infrastructure and technology, and process. The framework is designed in such a way that it can be adapted to other local conditions by changing the local parameters and data for whatever the regional case. In future studies, the proposed framework will be modelled using a computer program, i.e., i think, with a stockflow diagram simulating different scenarios. The cost-benefit changes for each scenario are to provide rational options among the simulation plans for decision-makers, city planners and other stakeholders, and to help predict future waste management needs.

REFERENCES

1. Su J, Xi BD, Liu HL, Jiang YH, Warith MA (2008) An inexact multi-objective dynamic model and its application in China for the management of municipal solid waste Management 28: 2532-2541.
2. Kollikkathara N, Feng H, Stern E (2009) A purview of waste management evolution: Special emphasis on USA. Waste Management 29: 974-985.
3. Kollikkathara N, Feng H, Yu DL (2010) A system dynamic modelling approach for evaluating municipal solid waste generation landfill capacity and related cost management issues. Waste Management 30: 2194-2203.
4. Zaman AU, Lehmann S (2013) The Zero Waste Index: A Performance Measurement Tool for Waste Management Systems in a Zero Waste City. Journal of Cleaner Production. 50: 123-132.
5. City of Austin (2008) Zero waste strategic plan the zero waste economy Prepared by Gary Liss & Associates, 4395 Gold Trail Way, Loomis, CA 95650-8929, USA.
6. ACT Government (1996) A Waste management strategy for Canberra- No waste Canberra.
7. Greyson J (2007) An economic instrument for zero waste, economic growth and sustainability, Journal of Cleaner Production 15: 1382-1390.
8. Seadon, JK (2010) Sustainable waste management systems. Journal of Cleaner Production 18: 1639-1651.
9. Kytzia S, Nathani C (2004) Bridging the gap to economic analysis: economic tools for industrial ecology. Prog Ind Ecol 1: 143–164.
10. Ossenbruggen PJ, Ossenbruggen PC (1992) Swap: A computer package for solid waste management
11. Computers, Environment and Urban Systems, 16: 83-100.
12. Chung SS, Poon CS (1996) Evaluating waste management alternatives by the multiple criteria approach Resources, Conservation and Recycling 17: 189-210.
13. Bovea, MD, Powell JC, Gallardo A, Capuz-Rizo SF (2007) The role played by environmental factors in the integration of a transfer station in a municipal solid waste management system. Waste Management 27: 545-553.
14. Beigl P, Lebersorger S, Salhofer S (2008) Modelling municipal solid waste generation. A review
15. Waste Management 28: 200-214.
16. Lu WS, Yuan HP (2011) A framework for understanding waste management studies in construction
17. Waste Management 31: 1252-1260
18. Chang NB, Davila E (2007) Minimax regret optimization analysis for a regional solid waste management system. Waste Management 27: 820-832.
19. Pires A, Martinho G, Chang NB (2011) Solid waste management in European countries A review of systems analysis techniques. Journal of Environmental Management 92: 1033-1050.
20. Forrester JW (1958) Industrial dynamics a major breakthrough for decision makers. Harvard Business Review 36: 37-66.

21. Sufian MA, Bala BK (2007) Modelling of urban solid waste management system: The case of Dhaka city Waste Management 27: 858–868.
22. Forrester JW (1968) Principals of systems Cambridge USA Productivity Press.
23. Bala BK (1998) Energy and Environment: Modelling and Simulation Nova Science Publisher New York.
24. Bala BK (1999) Principles of System Dynamics. Agrotech Publishing Academy Udaipur India.
25. Forrester JW (1971) World Dynamics. Wright-Allen Press MIT Massachusetts.
26. Meadows DH, Meadows DL, Randers J (1992) Beyond Limits Chelsea Green Publishing Vermont.
27. Bach NL, Saeed K (1992) Food self-sufficiency in Vietnam: a search for a viable solution System Dynamics Review 8: 129–148.
28. Saeed K (1994) Development Planning and Policy Design: A System Dynamics Approach Chelsea Green Publishing Vermont.
29. Saysel AK, Barlas Y, Yenigun O (2002) Environmental sustainability in an agricultural development project a system dynamics approach. Journal of Environmental Management 64: 247–260.
30. Mashayekhi AN (1990) Rangelands destruction under population growth: The case of Iran. System Dynamics Review 6: 167–193.
31. Sudhir V, Srinivasan G, Muraleedharan VR (1997) Planning for sustainable solid waste management in urban India. System Dynamics Review 13: 223–246.
32. Talyan V, Dahiya RP, Anand S, Sreekrishnan TR (2007) Quantification of methane emission from municipal solid waste disposal in Delhi. Resource Conservation & Recycling 50: 240-259.
33. Yuan HP, Shen LY, Hao JJL, Lu WS (2011) A model for cost-benefit analysis of construction and demolition waste management throughout the waste chain. Resources Conservation and Recycling 55: 604-612.
34. Yuan HP (2012) A model for evaluating the social performance of construction waste management. Waste Management 32: 1218-1228.
35. Thirumurthy AM (1992) Environmental facilities and urban development: a system dynamics model for developing countries New Delhi India Academic Foundation.
36. Mashayekhi AN (1993) Transition in New York State solid waste system a dynamic analysis. Sys Dynam Rev 9: 23–48.
37. Karavezyris V, Timpe K, Marzi R (2002) Application of system dynamics and fuzzy logic to forecasting of municipal solid waste. Math Comput Simul 60: 149–158.
38. Ulli-Beer S (2003) Dynamic interactions between citizen choice and preference and public policy initiatives a System Dynamics model of recycling dynamics in a typical Swiss locality.
39. Dyson B, Chang NB (2004) Forecasting municipal solid waste generation in a fast-growing urban region with system dynamics modelling. Waste Manage 25: 669–679.
40. Duran X, Lenihan H, ORegan B (2006) A model for assessing the economic viability of construction and demolition waste recycling the case of Ireland Resources Conservation and Recycling 46: 302-320.

41. Rehan R, Knight MA, Haas CT, Unger AJA (2011) Application of system dynamics for developing financially self-sustaining management policies for water and wastewater systems. Water Research 45: 4737-4750.
42. Yuan HP, Chini AR, Lu YJ, Shen LY (2012) A dynamic model for assessing the effects of management strategies on the reduction of construction and demolition waste. Waste Management 32: 521-53.
43. Goddard HC (1995) The benefits and costs of alternative solid waste management policies. Resour Conserv 13: 183–213.
44. Nordhaus WD (1992) The ecology of markets. Proc Natl Acad Sci USA 89: 843–850.
45. Farel R, Yannou B, Ghaffari A, Leroy Y (2013) A cost and benefit analysis of future end-of-life vehicle glazing recycling in France. A systematic approach Resources Conservation and Recycling 74: 54-65.
46. Gramlich EM (1981) Benefit-cost analysis of government programs Englewood Cliffs N J. Prentice-Hall.
47. Cowen T (1998) Using cost-benefit analysis to review regulation, New Zealand Business Roundtable
48. Posner RA (2000) Cost-benefit analysis definition justification and comment on conference paper. Journal of Legal Studies 29: 1153-1177.
49. EB (2008) Policy and consultation papers: a proposal on the mandatory implementation of the building energy codes
50. Thomas JM (2007) Environmental economics applications policy and theory Mason Ohio: South-Western.

AUTHOR NOTES

CHAPTER 2

Acknowledgments

The authors would like to thank the Social Science and Humanities Research Council of Canada for grant funding to both William Rees and Jennie Moore in support of this project. We also thank the Foreign Affairs and International Trade Canada (DFAIT)—"Understanding Canada-Canadian Studies Program" fellowship, and to the European Union ERG grant on urban sustainability to Meidad Kissinger. Finally the authors also thank Zoe Wang, for her review of Chinese journals.

Conflicts of Interests

The authors declare no conflict of interest.

CHAPTER 3

Acknowledgments

R. D. Arancon thanks the Department of Chemistry of the Ateneo de Manila University in the Philippines for the wonderful opportunity to learn. Also, heartfelt thanks are due to Jhon Ralph Enterina (University of Alberta, Canada) and Jurgen Sanes (Simon Fraser University, Canada) for help with some articles. Carol Sze Ki LIN acknowledges the Biomass funding from the Ability R&D Energy Research Centre (AERC) at the School of Energy and Environment in the City University of Hong Kong. The authors are also grateful to the donation from the Coffee Concept (Hong Kong) Ltd. for the "Care for Our Planet" campaign, as well as a grant from the City University of Hong Kong (Project No. 7200248).

C. S. K. Lin acknowledges the Industrial Technology Funding from the Innovation and Technology Commission (ITS/323/11) in Hong Kong. R. Luque gratefully acknowledges the Spanish MICINN for financial support via the concession of a RyC contract (ref: RYC–2009–04199) and funding under project CTQ2011–28954-C02-02. Consejeria de Ciencia e Innovacion, Junta de Andalucia is also gratefully acknowledged for funding project P10-FQM-6711. R. Luque is also indebted to Guohua Chen, the Department of Chemical and Biomolecular Engineering (CBME) and HKUST for the provision of a visiting professorship as Distinguished Engineering Fellow.

Conflicts of Interests

The authors declare no conflict of interest.

CHAPTER 4

Acknowledgments

The authors wish to thank the Government of the province Congo Central, the DRC, which initiated this research. We are grateful to Dr. Zinaida Zugman for her helpful advice on models used in this article.

CHAPTER 5

Acknowledgments

We hereby acknowledge the assistance of Niger State Environmental Protection Agency (NISEPA) and Niger State Ministry of Environment in preparing this paper.

CHAPTER 6

Acknowledgments

The authors would like to acknowledge the Municipality of Santa Cruz for its efforts in the management of waste and the data provided. We also like to acknowledge World Wildlife Fund and Toyota for their financial support to the waste management system in Santa Cruz Island.

Conflicts of Interests

The authors declare no conflict of interest.

Author Contributions

Marco Ragazzi and Elena Cristina Rada designed the research; Riccardo Catellani performed the research and preliminary analyzed the data under the local supervision of Xavier Salazar-Valenzuela; Vincenzo Torretta wrote the paper. All authors contributed to a deeper data analysis, read and approved the final manuscript.

CHAPTER 8

Acknowledgments

This work was funded by the Kaikōura District Council and the sustainable initiatives fund. The authors would like to thank Julia Schuler (Hochschule für Forstwirschaft Rottenburg) and Jeff Seadon (Scion) for their help in completing this manuscript, and all the stakeholders for participating in the hui in December 2011.

Conflicts of Interests

The authors declare no conflict of interest.

CHAPTER 9

Acknowledgments

This paper summarizes some of the results from the six-year research program “Towards sustainable waste management” [19]. Besides the authors a large number of people have contributed, including researchers, members of reference groups and participants in workshops and conferences. We are grateful to all these as well as to the Swedish Environmental Protection Agency for funding this research. Also comments from anonymous reviewers are appreciated.

Conflicts of Interests

The authors declare no conflict of interest.

CHAPTER 10

Acknowledgments

This article was supported by the Zero Waste Research Centre for Sustainable Design and Behaviour (sd+b) and the China–Australia Centre for Sustainable Urban Development (CAC_SUD) in the University of South Australia. The study is a part of an ongoing collaborative research project funded through the Co-operative Research Centre (CRC) for Low Carbon Living involving researchers from the University of South Australia, the University of New South Wales, the CSIRO and various government bodies. Special thanks to the Endeavour Research Fellowship Program for the support.

INDEX

R

S

X

Y

Z